Bill Oddie's
Gripping
Yarns

Bill Oddie's
Gripping Yarns

Tales of
Birds
& Birding

CHRISTOPHER HELM

A & C BLACK · LONDON

© 2000 Bill Oddie

Reprinted 2000, 2001

Christopher Helm Ltd, an imprint of A & C Black
(Publishers) Ltd, 37 Soho Square, London W1D 3QZ

ISBN 0-7136-5268-3

A CIP catalogue record for this book is available from the British
Library

Printed and bound in Great Britain by Creative Print and Design
(Wales) Ebbw Vale

CONTENTS

BIRD RACING

BITS AND PIECES

INTRODUCTION

I never meant to do it. Write a monthly column that is. Well, maybe a column wouldn't have been so bad. A short thin one. But a whole page? I knew I wouldn't be able to keep it up. After all, I'd already tried writing so-called humorous fortnightly articles for my local newspaper and been judged a failure on all counts, my offerings being neither humorous (I kept getting all serious) nor fortnightly (sometimes five or six weeks went by), nor even barely definable as 'articles' (jottings, more like). So how come I allowed Dominic Mitchell (the Editor of *Birdwatch* magazine) to persuade me to crawl under a permanently dangling Sword of Damocles and condemn myself to dwell in everlasting deadline hell? And, even more remarkable, how on earth have I lasted for well over five years and never once failed to deliver? And, more astonishing still, why do I hereby find myself having to confess that I have really truly *enjoyed* writing every one of these 60-odd pieces?

Heaven knows, I'd love to take the credit. But I can't. I honestly believe the reason that writing about birds, birding and birders is great fun, is quite simply because they are too. And if the articles entertain, it's because the subject does too. What's more, it is pretty inexhaustible. But I'm not. Which is why I was wearily apprehensive when the publishers approached me about producing a new book, and almost deliriously relieved when they suggested that it should be a collection of my monthly columns. Not that there wasn't work involved. I had to read them all again. And I enjoyed that too! I hope you do.

In fact, I have also done a bit of judicious editing — or rather, reinstatement. Quite often the layout of the magazine dictated that a few, sometimes quite a lot, of words had to go. Well, I've put them back. So what you have here are the original versions, uncut. I have also done a lot of reorganisation. Rather than simply leave the pieces in the order in which they were written over the years, I have collected them under various headings, and added a little 'linking material' when I felt it would clarify things further. (I can't just leave well alone can I?)

LOOKING BACK

Although the *Birdwatch* column has for some years been called Gripping Yarns, it in fact started life as Looking Back (arguably because I was the only contributor considered old enough to have an authentic past life; which I'm not sure was entirely a compliment, but never mind!). In any event, these first three pieces do exactly that: look back. So let's start at the very beginning (which, as Julie Andrews assures us, is a very good place to start) with a confessional so full and elaborate that it could not be contained within the pages of *Birdwatch* and therefore appeared originally in *BBC Wildlife* magazine. What's more, I *know* Julie wouldn't approve...

I was a juvenile egg robber
Confessions of a pre-teen oologist

First you do the work, then you have to promote it. Having recently finished a book and a new TV series. I am now embroiled almost daily in interviews on radio and telly. Nearly all of them have started the same way: "Bill, you're probably Britain's best-known twitcher...so tell us, how did you start birdwatching?" I then spend several minutes explaining exactly what a twitcher is and denying that I am one anyway, by which time the interview is over, and I haven't answered the question at all. So here's my chance. I'm going to skip the whole ghastly twitching bit and — at last — indulge in a bit of nostalgia.

Flashback to the late 1940s, the years of nutty slack and ration books. I was seven years old. I lived in Rochdale, Lancashire. I haven't been there for donkey's years (as we bluff northerners say), but back int' old days, when I was a lad, it looked just like it should have done: cobbled streets, flat caps, shawls and clogs, brass bands and black puddings. Remember the Hovis commercials? Just like that. Home was a terraced house in Sparthbottom's Road (by 'eck, they don't name 'em like that any more). I remember it well.

The one thing I don't remember is seeing any birds...except for one. I had my first ornithological experience under a privet hedge just in front of our house. I'd been kicking a ball around (a ball? Nay, a tin can at best, or was it a bundle of rags, or maybe a house brick?). Anyway, whatever it was that I was kicking rolled under the hedge. I crawled in to rescue it, and suddenly realised I was being watched by a pair of beady little eyes. It was a Dunnock, though I didn't know it at the time. All I knew was that it was sitting on a nest. I also knew that in that nest there were probably eggs. Obviously, the right thing to do would have been to crawl away and leave the bird in peace. But I was a schoolboy. I did what any 1940s schoolboy would have done: I coughed loudly — accidentally on purpose — and the bird flew off, allowing me to reach up into the nest. I still feel guilty at the frisson I felt as I also felt four eggs. I took one out and stared at it. It was a startling turquoise blue, arguably my favourite colour. I had to have it.

Thus began my life of crime. My dad made it worse. Shortly after, we moved to Birmingham (I was presumably deported from Rochdale for Dunnock abuse), and he bought me a copy of *The Observer's Book of Birds' Eggs*, to go with my *Observer's Book of Birds*. The fact that I could now match the egg to the species — small and turquoise = Dunnock (or Hedge Sparrow, as it was known in those days) — simply encouraged me to carry on birdsnesting. Okay, let's cut the euphemism: egg collecting. To be fair, it was something that every schoolboy of my generation almost 'had' to do, as obligatory a proof of impending manliness as scrumping apples or pulling girls' pigtails. But I knew it was wrong. I knew it then and — even more so — I know it now. And yes, I am ashamed of it. I shouldn't be admitting it, not even to myself. I ought to have blanked it. I should be in denial. The hobby that dare not speak its name. And yet...however...but...but...but...

The honest truth is, if I hadn't been an egg collector, I very much doubt if I would have become a birdwatcher. What's more — and this is even more contentious, and I know I'll get letters, but at least finish reading before you start writing — it was my egg-collecting experiences that taught me all sorts of skills and techniques. I actually learnt an awful lot. Of course, this isn't a justification, but it is a fact.

The first thing that rapidly impinged on my delinquent consciousness (if not conscience) was the variety of materials that birds use to build their nests. I learnt to recognise the species even if there was no bird nearby or no eggs to take. The memories are vivid. I can close my eyes and still see my first Chaffinch's nest, made from soft, grey-green moss and lichen, so that it looked as if it had almost grown naturally from within the fork of a tree. I learnt how apparently very similar nests could be distinguished by their different linings. Blackbirds line with grass, Song Thrushes with smooth mud, Chiff-chaffs with dry leaves, Willow Warblers with feathers. Coots build nests with sticks that usually float, Moorhens with dead reeds that tend to be drier.

I also began to appreciate how addicted some birds are to particular habitats or even specific plants, and thus I even acquired some very basic botany. I knew gorse when I saw it, and I also knew that was where I'd find Yellowhammer and Linnet nests. I knew nettles when I was stung by them, which I'd have to suffer if I wanted to get anywhere near a Whitethroat. And Spotted Flycatchers would almost certainly build their nests in an old wall covered in ivy, and so I'd have to risk a farmer's wrath by trespassing in his front garden if I wanted to check out his outhouse: and once I'd got that far, I might as well scuttle inside and shin up to the beams where Swallows were most likely to be nesting.

I got to know birds' songs and calls, and even to interpret them insofar as they were useful in my searches. If the Willow Warbler was singing, chances are it still hadn't found a mate, and so there wouldn't be a nest nearby. If, however, it was threading its way silently through the branches, it may well be on its way from the treetops to its nest, which was — rather surprisingly — at ground level, hidden deep in the grass. In which case, I'd wait to see where it dropped down. Or if it was giving its soft *hooeet* call, then it probably had youngsters, and so there was no point in my lingering any longer. Except that I often did, because the fact was that I was becoming increasingly fascinated by what the birds were doing. I stayed to watch the warbler parents catching insects or collecting caterpillars, and I watched as they approached the nest — usually by the same circuitous route — and I couldn't resist creeping forwards and counting the gaping mouths of the chicks, at this stage newly hatched and still unfeathered. Over the following days or even weeks, I'd return to that nest to see how the family was getting on, until the morning I actually saw them flutter feebly out of the grass and up into the supposed safety of a nearby hawthorn. I remember feeling terribly worried that they might end up as magpie or cat food — which was ironic, considering that, had I found the nest early enough, I would have happily aborted at least one little life by taking an egg.

And what about the eggs? Well, yes, I learnt about them, too. I browsed through my *Observer's* book and discovered that many hole-nesting birds have white eggs, presumably so that they can be seen more easily in the dark or maybe because pigmentation would be a waste of nature's time, effort and materials. The holenesters were out of my reach, but others weren't, and my collection grew.

It wasn't just a matter of acquisition. I often gazed at them and marvelled. I never ceased to be intrigued and impressed by the subtlety of the markings: blotches, spots, squiggles, colours, characteristic of particular species, and yet no two the same. I even had my favourite: the Yellowhammer, that looked as if it had doodled all over its egg with a particularly scratchy old pen, thus earning it the nickname of The Scribbler.

And how about the thrill and astonishment of finding my first Cuckoo's egg? It was, as it happens, in the nest of a Dunnock, alongside four more of those turquoise beauties. But it was off-white with dark brown spots, so obvious and different that at first I didn't realise what it was. Surely those devious Cuckoos would come up with something cleverer than that? Wouldn't they make some attempt at an accurate forgery? Apparently not. At least when I took that one, I was saving the lives of four Hedge sparrows...

But no. Hold everything. There I go again, almost justifying my delinquency. Yes, it's true, in my case, oology did lead onto ornithology — egg collecting turned me on to birds and other wildlife. It taught me the virtue of patience, it gave me an obsessive love of the outdoors, and I'm willing to bet that many other birders of my generation started the same way. But I really don't want to make out that the 1940s were some kind of golden age of nature-loving kids. There were far too many — er, how do I put this? — less *sensitive* boys whose interest was no more constructive than 'ragging' nests and batting the eggs around as if they were ping-pong balls.

So, let's get this straight, once and for all: EGG COLLECTING IS WRONG. Nowadays, it's illegal. Fortunately, it is almost unknown among modern schoolboys, and I trust any youngsters reading this piece will not be remotely tempted to follow my dreadful example. But there are, sadly, and almost unbelievably, still grown men — it's always men, not women — who do collect eggs, and an appalling destructive activity it is, especially because they are obsessed with rarities. Logically, therefore, it is almost in their interests to reduce a species to extinction: the last egg would be the most valuable. I repeat, egg collecting is a bad, bad thing, and like I said, I know it now, and I knew it then. But I really was a bit addicted. I needed aversion therapy. One day I got it. I was watching those Willow Warblers at the edge of the local golf course when I realised I was almost sitting in a Pheasant's nest. But I hadn't seen a Pheasant in that area for weeks — and, no, I hadn't sat on *them*, too; maybe they'd been felled by wayward tee shots. In any case, it was obvious from the state of the nest that the adult birds were now no more. The eggs were cold and wet and no doubt addled. I therefore felt justified in breaking my egg collecting rule and took two. I had no idea how long they'd been abandoned, but I was about to find out.

I took them back to our kitchen and began to attempt to blow them. I pierced a small hole in one end and a bigger one in the other. I huffed and puffed, but nothing came out, except the most awful stink. The stuff inside had turned into something the consistency of fish glue, and it smelt even worse. But I was not going to be beaten. If I couldn't blow it out, I was just going to have to...suck it out. I took the egg in one hand, my nose in the other, and a great dollop of appalling gunge into my mouth. I spat the gunge into the sink and placed the egg into a cotton wool compartment labelled 'Pheasant'. Then I took the second egg and did it again. Finally, I stood back to admire my collection and threw up over the whole lot.

And that was the day I gave up egg collecting and became a proper birdwatcher. But what is 'proper' birdwatching? Ah, well — that's a big question requiring a long answer, which is why I've already taken three TV series trying to convey what it's all about (and we still haven't done a programme on 'twitching'!). As for nests and eggs...well, by and large, the rule is: keep away from them, and thank the lord — or the BBC Natural History Unit — that we can now see such wondrous things on our television screens.

Of course, when I were a lad, we didn't have such fancy, new-fangled things as television sets. Well, that's my excuse.

'You are old, Father William'

Those were the days

In last month's *Birdwatch*, Anthony McGeehan referred to his group as 'having an average age approaching that of Bill Oddie.' I don't know if that was meant to be an insult (to me, or the group?) or a compliment, or maybe just a fact. I do suspect that the cheeky young whippersnapper (who must, incidentally, be knocking on 40 by now) was being a teensie bit satirical at my expense, if only because it's not the first time he's alluded to my age. I seem to recall him suggesting that the photo that heads my column shows a fresh-faced youth when I am in fact 'as old as Methuselah'. Well, just to prove that I'm not sensitive (much) about my advancing years, I hereby confess that I am 57. I don't have access to the birth certificates of Anthony's group, but Methuselah was reputedly 900. I must therefore suspect that though Anthony may be an undisputed dab hand at ageing Thayer's Gulls and Soft-plumaged Petrels he may not be quite so good at people. Never mind, I thank him. Not for reminding me of the inexorable passing of time, but for giving me an idea for this month's piece.

The idea is to consider how birding has changed since I was a 'youngster'. My thoughts will necessarily be brief and random, because I only have a page, whilst there's surely a book in this one. (In fact, I may well already have written it. You do tend to forget things when you get to my age.) I'd like younger readers (you know who you are) to read this and appreciate just how much we old fellas had to do without. Deprived, that's what we were. The rest of you can nod sagely and share some memories.

OK then, flash back to the late 1950s early 1960s and compare with now.

Bird Magazines: Only one. No *Birdwatch*. No *Birding World*. No *Bird watching*. Just *BB*. Which is why my quote in the ad —

'You can't call yourself a real birdwatcher unless you read *British Birds*' — *was* true when I said it! Spotters of gross anachronisms (Anthony?) will be pleased, or thwarted, by the fact that the new *BB* ad will carry an updated blurb and — at last! — a photo that doesn't make me look like an extra from *Woodstock* (youngsters ask your dads).

Bird Books: Not a lot. Especially when it came to field guides. The first edition of the British and European 'Peterson' was published in 1954 (not *that* long ago really. Well, not to me!) Before that, we carried around the *Observer's Book of Birds* (extremely portable, because it only had about 60 species in it!) and/or a pocketful of James Fisher's little *Penguins*. (books, not birds or biscuits.) Nowadays, there are more bird books than birds.

Opticals: Binoculars haven't actually changed all that much. I still have an antique pair of 8 x 30 Barr and Strouds that would give my snazzy little Leicas a run for their money (they'd lose of course, but not by much). Telescopes, however, have evolved incredibly. A modern compact lightweight spotting scope bears about as much relationship to the ancient brass draw-tube jobs as a Black-headed Wagtail to *Archeopteryx*. Believe me, you molly-coddled, tripod-toting young softies have absolutely no idea of the pain we had to go through. Imagine it, lying flat on your back for hours on end, neck unsupported (or with a rock for a pillow — if you were lucky) balancing a three-foot cylindrical monster on your knees (he's talking about his telescope, madam). Can you conceive how excruciatingly agonising that was? If not, you might care to try it sometime. If you can't find an old brass telescope, a large cucumber will do (and, come to think of it, it will probably be about as effective optically). And don't just do it for a few minutes, stay there for several hours. Then try and get up. I promise you, you'll have a new-found respect for the legendary old seawatchers who used to lie there for days staring at nothing off Selsey Bill. Then again, you might realise that they weren't so much dedicated as stuck. People died in that position and no one realised it for months. If you did survive, you were left with a posture like the Hunchback of Nôtre Dame. But we wore our stoop with pride.

Information technology: Minimal. If you are into the present day twitching scene, imagine this: no pagers, no birdlines, no mobile phones to talk you in. No 'Park in Tesco's multi-story, pay at the *Birding World* temporary turnstyle, wait in the queue, crawl through Lee Evans legs, and the bird will usually be found perched on the end of Steve Young's 600mm lens. Slides, videos, and signed photos printed while-u-wait.' None of that. There *was* a sort of 'grapevine' that operated amongst a few dozen top birders, but that was a very elite affair.

Then, sometime in the late 1970s, or was it 1980s?, Nancy's Café in Cley became the unofficial 'central office of information' for the latest sightings. The main flaw, however, was that if there was anything good around, the line was permanently engaged. If you could get through, you could bet there was bugger all about! For a long time the most up-to-date gen available was the somewhat ironically named 'Recent Reports and News' in *BB*. The definition of 'recent' being 'some time in the last month or two'. So how on earth did we old codgers add new stuff to our lists? We had to go and find our own birds. That's how tough *we* had it!

I'm going to have to stop there. Not because I'm getting upset by all the traumatic recollections, but because I'm not allowed any more space. I still think there's a book in this one. Or at least a few letters. So come on, put pen to paper (or is it finger to e-mail these days?) and tell us how *you* think things have changed. And have the changes been good or bad?

My own final thought is this: *I* love memories. It's just a pity you have to be old to have them!

Deeds of darkness
A September night and some strange goings-on

The other week I was talking Fair Isle with a relatively young birder. He was extolling the virtues of the observatory building. It is often referred to as the "birdwatcher's Hilton" and it *is* wonderful: light and airy and with huge picture windows, through which you could pick out a passing Daurian Starling a mile away.

"Yes, it's very nice," I agreed, "but, of course, it wasn't always like that." I am now at the age when I can't resist rattling on about how much we suffered when I was a lad and, as I reminisced, I realised how many of my early memories of Fair Isle were literally dim. For a start, the dilapidated RAF huts that were the old 'obs' not only had teenie weenie windows, but what glass there was had been thoroughly 'smoked' by the oil lamps that provided the only lighting. A Daurian Starling would have had to wipe the window with a damp rag and pull funny faces to get itself noticed.

By day, it was a strange twilight world in there. At night, it was positively spooky, as we huddled round a huge steaming vat of cocoa, like witches round a cauldron, and conducted the ritual log-calling, which usually had to be done again next morning as most of the ticks were in the wrong boxes and the writing was all squiggly.

Considering how gloomy the old obs was, you'd think I would have craved the daylight. Yet, strangely, many of my fondest birdy memories from Fair Isle are nocturnal. I shall never forget a September night spent up the south lighthouse. It was moonless, misty and wet: exactly the conditions when birds were drawn to the light like moths to a flame.

Even the journey down the island was exciting: a white knuckle ride on the back of a motorbike that would have been a major attraction at Alton Towers. The adrenaline pumped even harder as we scaled the lighthouse stairs and emerged on to the balcony.

The weather swirled around us. So did the birds. It was at the same time thrilling and slightly disturbing, as ghostly shapes flitted through the beams, some escaping into the gloom, others being drawn in by the lamp, fluttering frantically against the glass. Wheatears, pipits warblers; we caught so many we had to take our socks off and use them as emergency bird bags! It might have been a bit whiffy in there but I suppose they were the lucky ones, as we found several sad little corpses under the light in the morning.

The next time I went to Fair Isle the lighthouse had been floodlit and, though I was happy for the birds, I can't pretend I wasn't a little disappointed.

Fortunately, though, I then discovered 'dazzling'. For a good laugh there's nothing to beat it! It was a foul October night when a truly incompetent trio set out after a rather too hearty evening meal at the obs. There was me — a qualified ringer back then — my old birding chum Andrew Lowe, and the 'daddy of all twitchers', Ron Johns. Ron may be deadly serious in pursuit of a tick, but he is an equally avid seeker of a silly experience. He was not to be disappointed.

We were fully armed with a big butterfly net and a couple of strong torches but, so utterly black was the night, we could hardly see our own feet, let alone any birds. Nevertheless, we successfully dazzled three sheep, a large rock, and a gate post, before realising that we were totally lost. In fact, we had wandered into a particularly treacherous marshy area that we wouldn't normally have attempted to cross during the day, let alone in pitch darkness.

We should have panicked. Instead, we were rapidly reduced to helpless laughter. Andrew was the first to go, getting hysterical as he lost his wellies in the mud. I attempted to rescue him with the butterfly net, but only succeeded in tripping over a submerged stone, losing control of my balance and my bodily functions. My shock expressed itself in the form of a loud fart. At this, Ron, who was rather smugly just about to reach dry land, turned to make a ribald remark and stepped straight into a four-foot ditch. Well...maybe you had to be there! But I tell you, when I think of Fair Isle '76 that's what I remember, not the Pallas's Reed Bunting we found the next day.

RARITIES AND TWITCHING

I couldn't resist that Pallas's Reed Bunting reference there: the only species that I have ever been involved in adding to the British List. Since then, there have been a couple of other records, though it remains one of the rarest British birds, and undoubtedly one of the most boring. But who cares, it was a first!

Which brings me to rarities and twitching. Despite having written a whole book on the subject *(Bill Oddie's Little Black Bird Book*. Robson); I do not consider myself either a twitcher or an expert (it's still not a bad book though!). Neither am I anti twitching. But I have thoughts and feelings. And there's no better feeling than finding a rare bird, and no worse feeling than dipping out. (OK, I may not be a twitcher, but I can speak the language.) Hence there have been many pieces on the joys and heartaches of rare birds and those who pursue them...including me.

PG tips
A confession

Fair Isle's specialities: Lanceolated Warbler, Pechora Pipit, Pallas's Grasshopper Warbler, do occur elsewhere (especially in 1996, in places I wasn't), but if you want to give yourself better than lottery-winners' odds you really have to travel up to the magic — and very expensive thanks to British Airways — isle. The other thing they have in common is that they are all small, brown and skulky. There isn't a bona fide crippler amongst 'em. Nevertheless, they are all much desired birds. Especially by me.

Over the past four decades I have been on Fair Isle many times and failed to catch up with any of them. Chances are that if I had, I would have been in the company of a few other birders from the obs, and that would have been fine. The truth is, though — call me selfish or obsessed — that what I really dreamed of was finding one of the elusive trio for myself. Attempting to realise this dream, I've also spent many weeks on Out Skerries, a tiny island in the north-east of the Shetland group, and a mini Fair Isle, except that it doesn't get quite so many rarities (diminishing the chances of dream realisation) nor so many birders (increasing said chances). Anyway, the fact is that, despite many years of sporadic trying, I have failed almost entirely. Many a mile of turf I've tramped, but I have never yet trod on a Lancy, nor panicked a Pechora. But I said 'almost'. I did get one. I have never yet told the true tale...till now.

Date: 4 October 1983. I only have to re-read my notebook from that day to remind myself that it should have been good: 'Wind. south-east force 3. A lot of light drizzle developing into fog — or at least very thick mist — by afternoon.' It was my first day on the island, so I gave it my customary almost frantic thrash. Again I quote my notes: 'A fair basic selection of migrants: five Redstarts, one Whinchat, eight Goldcrests, five Blackcaps, four Robins, 20 Song Thrushes and a Redwing' — heart stopping stuff, eh? — 'There were also about 400 Snow Buntings. A lot.' I should say so! Considering I get excited about 10 these days. But despite the supposedly hopeful weather, it seemed nothing had really come in. Until late afternoon.

At about 4 pm, my optimism revived by coffee and a nap, I was clomping through what I laughably referred to as the Skerries 'reedbed', actually a small area of yellow flag, which looks quite pretty in bloom, and can occasionally fool marsh-loving birds into thinking they are safe from birders' boots in there. Of course, they're not, especially if the boots are on my feet. In gathering gloom, I flushed a single warbler. The visibility was dreadful. Murky and soggy. And so was I. So was the bird. All I noted was that it was a heavy-looking 'Reedy type'. It looked as dark as a Dunnock, and sported a rather large tail, but I couldn't even be sure if it was streaked or unstreaked. Thoughts of Savi's or River Warbler flashed through my mind. I kicked it out — er, I mean flushed — it again. This time I reckoned I did see streaking, and I also noted a paler rump as it plunged back into the 'reeds'. On the third flush, fortunately, 'it flew towards a stone wall, perched on a tangle of barbed wire, and clearly showed the characteristics of Pallas's Grasshopper; all of which I was able to confirm when I later caught it in the small Island heligoland trap further along the wall'. Well, that's what I wrote when I sent in the record to *British Birds*. The truth is a little less straightforward, and involves a confession of criminal activities.

What really happened was the rain increased, and the fog got thicker, to the point that I could hardly see the reedbed, let alone the bird. I was, however, convinced that I had something good. So I raced back to the wooden chalet that we — my wife Laura and I, that is — called our holiday home. I rummaged in my rucksack and produced a crumpled plastic bag. Inside the bag was an ancient — and somewhat holey — single-panelled mist net. OK, yes, I did once have a ringing licence, but I had let it lapse many years before. I had no rings...but I did have that net. I'd kept it for 'sentimental reasons', of course; oh, and just in case I ever found myself pursuing a mystery warbler in a stunted reedbed in almost pitch darkness. In pouring rain, Laura helped me rig up the net. Not the highlight of her holiday, I imagine. We caught first a Reed Bunting, which I took an instant dislike to because it wasn't a warbler, then we trapped a leaf, and on the third 'drive' we got the bird. I stuck it in a sock, which was the nearest thing I had to a bird bag.

Next step, back to the chalet. I had Svensson with me, what a happy coincidence, considering I had no intention of trapping birds, but all he told me was that the wing formula of *certhiola* is not distinct from *naevia*. Size is, of course, but I had no measuring instruments. (Not that well prepared.) Nevertheless, undertail-covert pattern and dark terminal tail band all pointed in a promising direction. But where were the legendary white tail tips? PG tips, that's the joke isn't it? The joke seemed to be that they weren't there. But it surely wasn't a giant 'normal' Gropper? Second opinion and better light required.

I scampered off across the island, through the ever worsening storm, clutching my sockful of potential dream realisation. Edwin lived by the quayside, still does, and he is a birder and — more to the point — a qualified ringer, fully equipped with a ruler, and a camera. Into the 'ringing room' we went: the rather dimly lit garage. I handed him the sock, and invited him to feel inside. He did. (Ringers know no fear.) "Mmm", he said, "it feels like a Blackcap." I assured him that it wasn't. "It's a Grasshopper Warbler." "Well, it's a bloody big one then." Quite so.

The processing began. Everything fitted: size, and every plumage detail. Except one. Well, several. Still no white tips on the tail feathers. We were both baffled. 'We'd better take a photo at least.' Edwin held up the bird. The flash flashed. And so did the warbler. Suddenly it spread its tail. On the outermost tips of the outermost feathers there was definitely dazzling white. Well, dirty pale grey really, but white enough to make a dream come true.

Why hadn't we noticed the tips before? I put it down to poor light, and over excitement, but, most of all, to the fact that the bird had kept its tail tightly clenched until it got in front of the camera.

What a performer, and prima donna-ish with it. "I'll show you my tips when I'm being photographed, and not before!' We felt honoured and elated. It was doubly satisfying when, a few days later, it performed again, this time genuinely on the barbed wire by the wall, on a glorious sunny Shetland afternoon, and I was able to share it with several local birders. Not a huge twitch, just some of the folks you somehow felt 'deserved' to see it...such as Iain Robertson, Nick Dymond, and the late Bobby Tullock. I know I'm being elitist, but sod it. If I ever find another one, I promise I'll put it on Birdline.

Meanwhile, I await the letter drumming me out of the BTO for illicit use of a single-panelled mist net.

They may take away my membership, but they'll never destroy my dream!

The big dipper
What is it about Shetland in spring?

What have I ever done to Shetland? Been nice, that's all. Over the years, I've poured money into the island economy, spending hundreds of pounds on extortionate air fares and hire cars. I've provided front page news for the Shetland Times by burning down a holiday chalet on Out Skerries. I raced up there to shed tears on the telly during the Braer disaster; and I was back again a few months later raving about the recovery of the islands on The Travel Show. In short, I have never missed an opportunity to be nice to Shetland. And yet how does it repay me? With mega dip-outs, that's how.

It began in mid-May 1975 when I drove unwittingly past a female Wilson's Phalarope bobbing about on a small loch that I didn't bother to scan until the next day, by which time the bird was no doubt half way back to Canada — unless it was hiding behind a Mallard laughing at me. After that, my 'spring dip-out' became something of a tradition. I was usually on my way to Skerries or Fair Isle, with a few hours to spare on Mainland. One year I missed a Green-winged Teal by half an hour. In 1984 it was a Needle-tailed Swift, and in 1985 a White-throated Sparrow: both 'you should have been here yesterday' birds.

And yet I still keep going back. I was there again for a week this May, the 20th anniversary of the phalarope dip. I should have known Shetland would have something particularly painful lined up for me! The first few days were truly record-breaking — for the total lack of migrants. "Worst spring we've ever known," muttered the dispirited locals. Nevertheless, I stomped round Skerries and at least turned up a Peregrine — a rare bird up there — doing its best to impersonate a Gyr by swooping through a snowstorm (yes, snow in late May!).

The following day was better. I flew to Mainland, ticked off a couple of 'resident' King Eiders and found a male Golden

Oriole in one of the few truly verdant gardens on Shetland. The poor but lovely bird sat there, posing among the celandines and wood anemones, wondering why its wing-tips were frozen together. But it was a sign of spring — at last — and it lifted my spirits no end.

I carried on up to Unst. The main purpose of my visit was to open the new visitor centre at Hermaness, at the very northern tip of the islands. It was a great day: a splendid event, a fabulous habitat and even dazzling sunny weather. Shetland at its most delightful. (There I go, being nice again.) If you haven't been to Hermaness you really should. Then again, if you have a single twitching gene, you've probably been to tick off Albert the albatross. Fortunately, as it turned out, I did exactly that in 1975 so I didn't actually 'need' him. It would have been nice to pay my respects again though...but no.

Displaying the impeccably cruel sense of timing that has taken him 20-odd years to perfect, Albert had flown off 10 minutes before we arrived above his outcrop. Never mind, I thought, enjoy the walk and the weather, and get back in time to watch the European Cup Final on the hotel telly.

So I set off. Shetland is big: just under 100 miles from the very north to the very south...from Hermaness to Sumburgh, in fact. I just made it. I skidded into the car park of the Sumburgh Hotel, scooted up the stairs and zapped on BBC1, even as the teams trotted out. Two hours later, Ajax had beaten Milan and I was satiated with the justice of it — and with cod and chips and crisp white wine, room service. Life was good. It could only be made better by a rare bird.

Right below my bedroom window was the famous Sumburgh Hotel garden. The trees are pathetically stunted, but it has a rarities list that almost rivals Fair Isle's. So, out for a quick 'thrash'. It was only when I fell down the steps that I realised it was nearly dark. Nevertheless, I managed to eke out a sleepy Whitethroat and one 'other bird' (of which more in a minute). No rarity, but never mind.

The next morning I scanned the garden again, along with a local birder. A Rosefinch sang sweetly, thus making up for its disgraceful appearance (a case for the Trades Descriptions Act, or what?) and a Bluethroat appeared and disappeared in the same movement, as they do. The Whitethroat was still there, but not the 'other bird'.

I was leaving in the afternoon, so I still had a morning's birding ahead. After bacon and eggs, I rang the airport to confirm my flight. "There's bad weather on the way," warned the lady from British Airways. "F.O.G. But we can get you on the morning plane." I had work to get back to and couldn't afford to be stranded. An hour or so later Shetland was disappearing into the gloom, but I was homeward bound. Not one of the best springs, I thought, but at least no major dip-out this time. Oh no? Not so fast, Bill.

That afternoon I sat in my London office, writing up my notes from the final site: Common Rosefinch, Bluethroat, Whitethroat and the 'other bird': "A 'phyllosc'. Seen only briefly, in fading light. Flitted away across nearby weedy patch. Looked brown and dingy. Presumably a Chiffchaff". Then I made that masochistic phone call that I suspect most birders punish themselves with after a trip. I dialled Birdline. First bird up: "In Shetland, an elusive Dusky Warbler, in the Sumburgh Hotel garden". Oops!

I've been trying to console myself ever since. Dusky Warbler is the most boring bird in the world...but it would have been a British tick! Anyway, I'm telling you, I'm not saying anything nice about Shetland ever again. Well, not till the next time.

Making a spectacled of myself?

The art of losing a bird with bad grace

OK, I have to admit my articles have been a bit frivolous lately. This month though I'm going to deal with a very serious matter: losing a bird. We've all had a few 'armchair ticks' in our time, but have you ever had an 'armchair loss?' I have. It's a very narking experience. And I'm not talking about 'lumping' either. No, I'm talking about a major rarity. Maybe you've got the species on your list. Filey, late May, last year? That's the one; Spectacled Warbler. Deemed to be Britain's "first acceptable record".

Mmm, interesting phrase that, isn't it? It surely implies that there've been previous 'unacceptable' records. Well, yes, indeed there have. But, more than that, there have been other records that were, in fact, accepted by BBRC (the Rarities Committee) for a varying amount of time. I can think of three straight away. There was a bird trapped at Spurn in October 1968 and for quite a few years accepted as Britain's first, until someone re-read the notes and quite rightly pronounced it an unambiguous Subalpine. Then there was a "small Sylvia" seen at Porthgwarra, October '79, also accepted as Spectacled for several years. And, finally, there was 'mine'. Allow me to tell the tale, albeit it's going to hurt me.

Fair Isle, 4 June 1979. There'd been several days of south-east winds. We'd already had Subalpine, Marsh and Icterines, Ortolan Bunting, several Grey-headed Wags and so on. Midday on 4th I decided to forgo lunch and make do with a Mars Bar, while gazing into the murky crevasses of the North Reeva. For those of you who've never been to Fair Isle, the Reevas are two huge natural 'holes' which provide excellent shelter for weary wind-blown migrants who find calm and insects on the rocks and seaweed of the beaches below. Spotting birds down there is surprisingly challenging as it's only when a warbler flits into view that you realise just how high the cliffs are.

The technique is to sit there patiently, with bins and scope poised, and give it time. This is what I did, and it worked. I quote from my notebook, written up that evening: "At about 13.00 hours I noticed the sudden appearance of a small number of Lesser Whitethroats in the south of the Island. As I watched the North Reeva from the cliff top I soon found one or two Lesser Whitethroats and what, for a brief moment, I assumed to be a third, until it settled, revealing bright chestnut panels on the wings. I watched it for half an hour, taking full field notes, and, when more or less convinced that it could only be a Spectacled Warbler, I called the obs. Ian Robertson (the warden) et al arrived and immediately confirmed the identification (he saw lots in Israel only a few months ago) and Nigel (assistant warden) is familiar with them from the south of France."

I admit I was extremely pleased with myself, especially as, at the time, I hadn't seen Spectacled Warbler elsewhere in Europe (hence the cautionary "more or less convinced"). We made several attempts to mist-net the bird but, not surprisingly, failed. Anyone who knows Fair Isle will confirm that mist-netting in the Reevas is not only unlikely to succeed but is rather a dangerous activity. Nevertheless, copious notes, sketches and paintings were compiled by all observers and — again not surprisingly — the record was duly accepted as Britain's third Spectacled Warbler. Yes folks, the Spurn and Porthgwarra birds had happily survived for 10 years.

So, what's happened since? Well, like I said, the Spurn bird was eventually demoted to Subalpine. The rather mysterious Porthgwarra bird was then equally mysteriously reassessed and rejected; and so was the one I like to think of as 'mine' (though no disrespect to the dozen or so highly competent observers who enjoyed it with me).

Naturally, when I heard I'd lost my Spec I was, let's say, a little upset. The next time I bumped into (or do I mean collared?) a bloke on the BBRC I enquired casually — yet tearfully — why they had changed their collective mind. I was gently, tactfully, but firmly informed that when the Spurn bird was discredited, the Porthgwarra bird became the 'first', but that it's credentials weren't good enough to be accepted as a new bird for Britain, so it has gone too. This meant promotion for the Fair Isle bird, but apparently this was not good enough to be a first either!

Well, what do we make of that? Do we conclude that the criteria for accepting seconds and thirds are somewhat laxer than for firsts? Maybe that's fair enough. Does the committee reason: "Oh well, it's occurred before, so it probably was one!" In which case...since the much-twitched and photographed Filey bird was indisputably a Spec, shouldn't the Porthgwarra and Fair Isle birds now be re-instated? and if not, why not? Heaven knows, I don't want to sink — or do I mean ascend? — to the depths/heights of sarcasm of Alan Vittery's recent open letter to the BBRC (*Birdwatch* 15) but I wouldn't mind asking — rhetorically, of course — what exactly does the Committee think that bird was that I found on Fair Isle, 4 June 1979, and watched along with lots of other people, at a time when there was a Roller and a Sardinian Warbler on Shetland and a singing Rüppell's on Lundy, and five days before Britain's second Cretzschmar's Bunting occurred on...guess where? Yup, Fair Isle. Well, I know what it was, 'coz I've seen hundreds of 'em since, from Cyprus to the Canaries (though I didn't go up to Filey last May) and it's still on my British list.

Look, I'm not being bitter (honest). I'm just having a little muse about this acceptance/rejection business. And the thing I tell myself, and Alan, and everyone else, is this: we all know, in our heart of hearts, if any of our rarities are stringy. And I'll tell you something else: the BBRC know that some of theirs are too. I'm not naming any names or species, but I'll stake my binoculars on the fact that there have been plenty of dodgy records accepted in the past (and no doubt will be in the future), and they might even include a 'first' or two! What's more, I'm absolutely certain that even more 'good records' get rejected! But I'm not knocking the BBRC. Honestly I'm not. It's a difficult job, and I wouldn't want to do it. But you know, I am beginning to wonder if it all matters quite as much as the letters, and arguments, and bad feelings would suggest. In fact...maybe I shouldn't have written this piece. Ah well, it's too late now!

PS. I do rather like *Dutch Birding*'s habit of giving reasons for rejections. That is fun.

Picture this

Some funny goings-on in the Netherlands

Do you get *Dutch Birding*? You should. It's an excellent publication, even if it does arrive in a rather over-discreet white envelope and the postman probably thinks it's a porno mag. Mind you, as it happens, there is a pretty sexy picture on this month's cover — of a Ross's Gull displaying to a Black-headed.

But the stuff inside is even naughtier. It is the annual 'Rare Birds in the Netherlands' issue. The systematic 'acceptances' list makes interesting reading, but it's the next bit that is really rather scandalous. *Dutch Birding* not only gives a list of birds not accepted, but also tells us why they've been rejected. You wouldn't get polite old *British Birds* doing that, eh? We're far too gentlemanly on this side of the North Sea. How does the British Birds Rarities Committee put it? "In the vast majority of cases, the record was not accepted because we were not convinced...that the identification was fully established." How discreet. "In only a very few cases were we satisfied that a mistake had been made." So at least tell us about them then! Why should the rarities committee have all the fun? They needn't name names, but they could at least publish 'mystery photos' of known stringers in funny positions — like holding their heads in embarrassment.

Actually the CDNA (Dutch rarities committee) doesn't go that far, but they do go into the sort of detail that must hurt somebody's feelings — as well as entertaining their readers! To be fair, there is a valid argument that explanations of rejections can be very instructive to anyone thinking of submitting a rarity. Here are some of the reasons *Dutch Birding* gives for not accepting records:

"Gull-billed Tern, 9 May. No description." Lesson: write a description.

"Dowitcher sp, 6 Nov. Description does not exclude Bar-tailed Godwit." Lesson: compare confusion species.

"Greenish Warbler, 26 May. CDNA considers five seconds too short for a positive identification." Lesson: lie about the time.

Other verdicts include the positively humiliating: "Siberian Tit. Spring. Two. Poor description. After checking by a member of CDNA it turned out to be a pair of Tree Sparrows." Oops! And hand that man a loaded pistol. Or there's the downright suspicious: "Short-toed Lark. 22 Oct. Description quite complete, apart from a call belonging to another species, but CDNA finds it hard to believe that this bird could not be relocated by some of the 200 birders searching for the Isabelline Shrike in the immediate vicinity." So what are they suggesting? That this guy misidentified the shrike as a lark? Or that the lark was doing impressions? Or did the 200 birders include several members of the CDNA who dipped out on the Short-toed? Mmm, intriguing stuff.

But not half as intriguing as this one: "Kittlitz's Plover. 30 April 1990. Identification accepted, but CDNA is not convinced that the photograph was actually taken at Den Helder, based on the extremely sharp contrast of the shade and the high position of the sun in the photograph, the fact that the location of the photograph could not be found and the fact that the grass species on the photograph would all be blooming extremely early in the year."

And perhaps the elephant lurking in the background was a bit of a give-away? And yet the statement continues: "Still, the CDNA does not question the integrity of the photographer." Just his sense of geography perhaps? "Apparently an incomprehensible exchange must have taken place; all in all, this was one of the records with the longest circulation in the existence of the CDNA."

Yes, indeed, five years, but now — at last — the guy has discovered where his Kenyan slides went! Or are they saying he really was suffering from the delusion that he'd taken that picture in Holland — and for five years they almost believed him?

Well, this is all a bit worrying. I mean, we all thought we could trust photographic evidence to support rare birds, didn't we? But no more. What a headache for the committees, especially if this sort of mix-up catches on over here. They'll have to learn how to judge the authenticity of photos as well as descriptions.

No doubt the BBRC will remain as polite as ever: "The image of the supposed Dupont's Lark is extremely blurred owing to excessive camera shake, suggesting that the photographer was completely out of his head at the time, probably on cheap Spanish brandy. This no doubt explains why he accidentally claimed that the picture was taken in mid-November on Hampstead Heath."

OK, accidents can happen. But what really bothers me is that we could have another potential Hastings Rarities syndrome on our hands. Can't you just imagine the more competitive members of the '400 Club' slipping in a few shots from their latest trip to Antarctica: "King Penguin. Two million. Hengistbury Head." And how will the committee decipher the deception? "The shadows cast by the 100-ft iceberg in the back of the shot look to be at the wrong angle for Dorset; moreover, we have been unable to trace any American oil exploration activity in Christchurch Harbour, and finally we find it hard to believe that two million penguins would not have been reported by at least one of the 300 holidaymakers also normally present at this location."

But these photographer chappies are cunning fellows. Have you heard of a 'slide sandwich'? Slap a couple of piccies together and — Bob's your uncle — you've got an Ostrich on St Agnes being admired by 2,000 twitchers. What's more, they'll all be so anxious for a tick, they'll swear it really happened. Sounds like a job for the CDNA to me.

Alive and ticking
Why you can't count a dead bird

Why these morbid thoughts? Well, I've just been musing on the somewhat bizarre scenario that occurred up in Shetland this winter when a local birder found a Brünnich's Guillemot along the shoreline.

It was alive, but clearly not well. But it was still 'tickable'. (You can tick poorly birds, as long as they're not in captivity, of course.) So, he no doubt ticked it. Then he took it into care, thus rendering it untickable. However, a number of English twitchers flew north, or put themselves on permanent standby, so that they could be present when the bird was — hopefully — released back into the wild.

Alas, the 'patient' did not make a rapid recovery. So, picture the scene. There, throughout the long, dark Shetland nights, poor old Brunnie lies in a coma, on a bed of locally-plucked Eider down while, gathered at his box side, the twitchers conduct their candle-lit vigil. In the glow of flickering flames, the anxiety on their faces becomes ever more anguished. In hushed tones and sensitive words they pray as only twitchers can: "Dear Lord, please let the little bastard recover quickly so I can tick him off...oh, and please make the bloody tattler reappear so I can get that on the way back...oh, and I need King Eider for my year list, too."

I'm not sure if the Lord complied with their requests. I know the tattler pushed off, but I'm not sure what happened to Brunnie. (I got bored listening to progress reports on Birdline and missed the final outcome). What I'd like to think happened was that the wretched bird eventually opened its eyes and squawked "Where am I?", at which the twitchers let out a mighty cheer and it instantly died of shock.

No doubt it was then given a ceremonial send-off in true Shetland fashion. They do these things rather well up there (it's that Nordic tradition, you know). They'd wait for a night when the aurora borealis was illuminating the sky. Then the bird's corpse would be pushed out to sea in a Viking long-boat, while ranks of weeping twitchers held blazing torches aloft and sang a mournful dirge.

Now this is the sort of ritual that American tourists would pay good dollars to witness. So, I really wouldn't blame the Shetland Tourist Board if they are arranging for a fridge full of Brünnich's Guillemots to be flown in from Iceland at this very moment. Heaven knows, I don't want to accuse the locals of opportunism but I tell you, if I were a long-distance twitcher, before I spent several hundred quid nipping off to the Northern Isles for another release, I'd ask to see cardiographs and pulse-rate charts to prove that the bird was actually alive — and I'd want it to sign a certificate vouching that it was truly wild.

Like I said, you can't tick dead birds. As for the 'morality' of ticking birds that have been extremely unwell, nurtured back to health, and then discharged from care at a pre-arranged and widely-advertised time and place...well, I'm not going to question twitchers' principles (not this month). But I will examine my own.

I have to confess that I did once tick dead birds — as long as I'd found them myself! This is, of course, a pretty dodgy qualification since how was I to know that the mangy old corpse hadn't been found by someone else and tossed away? But I was happy, as long as the setting was reasonably 'natural' (I drew the line at birds found in dustbins.) Thus, my first Sooty Shearwater was actually one wing found along the tideline near Cley; and my Woodcock tick referred to an unfortunate migrant that had crashed into my kitchen window during a foggy March night.

In both cases I had the excuse that I was very young at the time. In my more mature years, I learnt to be more rigorous, but this didn't extend to disqualifying an ex-Great Reed Warbler which a crofter on Out Skerries had picked up in his rhubarb patch. I remember he told me he'd left it on his back porch so I could check what it was. He thought it was a Yellow Wagtail!

It was about then that I further qualified corpse-ticking to 'it's okay if you are the first to identify it'! But eventually, inevitably, came the day when I finally accepted that you can't tick a dead bird (I think that it was the same day that I found a live Great Reed on Skerries).

While I'm in this macabre mood I'll end with one last corpse story. OK, you can't tick 'em, but they are fun to find. Especially when the circumstances are as unlikely as on 6 December 1990. I was visiting the RSPB headquarters at The Lodge in Bedfordshire. On my way in I called at the shop near the entrance. A young gardener grabbed me, saying that his colleague had just found a dead bird under a hedge. He thought he knew what it was, but would I please confirm it. I did. It was what he thought. An American Yellow Billed Cuckoo — just like the one in the drawing, only rather less alive!

The one that got away

Sometimes birds that 'don't seem right' come back to haunt you

U.F.O SNIPE...

overall rather dark wing

rounder wing tip?

dark undawing

rather indistinct scap-edges.

white belly.

BRYHER APRIL '83.

gtr covs barred buff.

no obvious white on tail

thin pale edge to gtr covs + 'Λ' at join with primaries.

v.thin trailing edge - hardly clear white - only visible at very close range

Never called. Often short straight flights.. or.. slow zig zag at most. Overall, rather large, bulky + slow....

The ones that got away. We've all had them. The half heard call or the fleeting glimpse that left you with the funny — and rather upsetting — feeling that you've just 'thrown away' a crippler. All a little galling, but they've never bothered me too much as — to be honest — I really didn't see any of them well enough to be tempted to tick them.

However...I do have one that hurts a little harder. Not only did I see this bird pretty well, several times, but it would have been a first for Britain. It still would. In fact, I have to thank *Birdwatch* for reminding me about it — thanks a lot! — in an article in the January issue: The Next New British Wader? If you don't instantly recall what the species was, hang about and hear my woeful tale.

Time and place: Scilly, Bryher, April 1983. I was staying at a guest house run by friends who live on the island. The bloke would not claim to be an expert birder but he does have a pair of Zeiss and a field guide and he's not daft. Let's face it, he also lives on Scilly and works outdoors every day. The fact is, he might not be too hot on tertial fringes and primary projections but he's likely to know something unusual when he sees it.

I had a couple of days of my week's holiday left when my friend popped his head into the breakfast room and announced: "Oh, by the way, Bill, I'd meant to mention... I think I keep flushing a Great Snipe from the far end of the pool." His nonchalance seemed almost obscene. "Keep flushing?" I asked. "How long has this been going on?" "Oh, the last two or three mornings. The dog keeps putting up this snipe. It's a bit like a small Woodcock." (Yes, he knows Woodcock.) "I'm sure it's not a normal Snipe." (And yes, he knows 'common' Snipe — or whatever it's called this week.) "It's got a shorter bill and it always flies straight and drops back down."

It sounded like a Great Snipe to me. But then what did I know? Well, back in April 1983, not as much as I know now! At that time Great Snipe was one of my 'top-of-the-list' bogey birds. I'd missed them by days, or even hours, over the length and breadth of the country, from Fair Isle to St Agnes and, heaven knows, I'd followed up quite a few stringy ones only to find undeniable *gallinago*. In fact, my search only ended this spring in Cyprus. When I did finally flush my Great Snipe it was a truly classic demonstration of the old adage 'when you see it, you'll know it'. But I didn't know it back in April '83. Or maybe I did.

Anyway, off I scampered to the north end of the Bryher Pool. I clomped around for half an hour or more in the relatively dry tussocky area. The 'right' habitat for Great Snipe, from what I'd been told. Nothing got up. This I took to be a very encouraging sign. A normal Snipe would surely have shot out zigzagging and screeching, so there had to be a Great Snipe in there somewhere. If only I could tread on it!

I calmed down and reminded myself of the clincher characteristics of Great Snipe. Firstly, they are invisible. Secondly, they are silent. I was doing well on the first point. I was about to score on the second. Down went my welly. Up went the snipe. It flew in a perfectly straight line and plummeted into another tussock, surely confirming its identity: invisible and silent. Yes, but what did it look like?

How had my friend described it? "A bit like a small Woodcock." Er, well, yes, it was a bit…a very little bit. In fact, more like a slightly overweight Snipe, I'd say. Mmm. I had to admit this wasn't the "you'll know it when you see it" certainty I'd anticipated.

I shall now cut a fairly long story short. I followed that bird for most of the penultimate day of my stay. I saw it on well over a dozen occasions. Mostly in brief flight views, but once on a more prolonged fly past. I never saw it "out in the open" on the ground, but I did get a great view of it as it pitched on the far side of the pool with wings wide apart and tail spread. In birding parlance I pretty well "burned it up". And, by the end of this scrutiny, I was absolutely certain that it was definitely not a Great Snipe. I was disappointed but, in a strange way, satisfied that I had done such a thorough and righteous investigation.

It was only on the next — and final — morning, when I was about to leave the island, that it occurred to me to ask myself the somewhat leading question: "OK then, if it wasn't a Great Snipe, what was it?" A few exotic names and a sense of panic raced through my head. My legs raced back to the pool. I managed to flush the bird for a last tantalisingly brief and disturbing view. Whatever it was...it didn't look "normal"!

I had got pages of notes and, back in London, I began to analyse them. They were mainly 'negatives' concerned with eliminating Great Snipe. "No obvious white on sides of tail"..."no white-edged 'panel' on the upper wing"..."no barring on the belly"..."no white trailing edge to the primaries and secondaries"...Er, wait a minute..."No white trailing edge!"

OK. If you haven't got there ages ago, you will have now. I quote that article from *Birdwatch*: "The potential vagrancy of Pintail Snipe (*Gallinago stenura*) to Britain has long been expected. Some 'possible' individuals have been claimed in the past." Like on Bryher in April 1983?

Well, in fact, I never did claim it. I did ring my friend on the island and ask him to confirm that the bird really didn't have a white trailing edge. His answer: "Definitely not." I did ask if he had ever heard it call. Answer: "Just once. Softer than a common Snipe. A bit lower, maybe...but different." I also rang up my mate the late David Hunt and asked him if he could track down the bird but, alas, his tour-leading schedule didn't include Bryher until the following week, by which time the UFO had disappeared. I corresponded with D I M Wallace, I rummaged through skins at the British Museum, and I pored over obscure papers from Kenya and the Far East. All signs led to stenura. I think I even suggested that the Rarities Committee put it on their files — "just for the record". But I didn't claim it then and I've no intention of claiming it now. Mind you, I must confess, as I was sorting out three species of snipe in a foul-smelling paddy field in Hong Kong a couple of years ago, I couldn't help muttering: "There goes the one that got away!"

A sting in the Bluetail
Why do twitchers twitch?

I just want to be alone!

I've just been re-reading the various accounts of the Bluetail mega-twitch at Winspit last autumn. Crowds of several hundreds (according to the bird magazines) or several thousands (according to the newspapers), shoulder-to-shoulder, rank upon rank (according to everyone), queuing patiently, or pushing and panicking (depending on when you were there, apparently), and even...birdwatchers bursting into tears!

Birders? Crying? I had to re-read that one several time before it sunk in. Surely this was some hack from The Sun exaggerating the whole thing. Nope. There it is, reliably reported in *Birdwatch* and *Birding World* and *Bird Watching* (and no doubt even in *British Birds*...eventually). Who'd have believed it? Blubbering birdwatchers, Batman!

Well now, let's imagine we didn't know that all this palaver was aimed at a little brown (with a bit of blue on it) bird. When we hear about crushes and crowd hysteria like this it's usually directed at...the Harrods sale, possibly? Manchester United, maybe? But no. What it sounds most like to me is the response of a girlie teenage audience at a Take That concert! Our youngest readers will be aware that Take That are a 'heart-throb' pop group. Slightly older readers may care to substitute Bros. Or maybe you go back to the Bay City Rollers. Or, if you're my generation, The Beatles. Or, if you really want to give your age away, how about Frankie Vaughan or Dickie Valentine? Anyway, these pop idols all have (or had) something in common with the Winspit Bluetail. They reduced their fans to tears.

So how far do birding groupies take this thing? Do they simply weep and wail at the unbearable agony of a potential dip out? (The equivalent of failing to get tickets for the concert or, worse still, mummy or daddy — or wife — telling you you can't go.) Or do they also burst into tears of joy at the sight of those sexy blue retrices?! Do they scream out: "Bluetail, we love you!"?

Maybe they tear off their underpants and hurl them at the bemused little chat. Do they try and snip off a feather as a treasured keepsake and keep it under their pillows? (Given half a chance, I bet they would.) And how about 'wannabes'? If you go to a David Bowie gig, everyone dresses up like their hero. So why don't twitchers? I reckon nobody should have been allowed into Winspit valley who wasn't wearing red flanks and a blue tail. Actually, if that had happened, I might even have been tempted to go and have a look myself. I might even have tried to push to the front and tick off the bird as well.

As it is, I still 'need' Bluetail. I am Bluetail 'blocked', or whatever the rather anal expression is. And it doesn't bother me one bit. Okay, call me a crotchety old spoilsport if you like (which I suppose I am) but I really can't help feeling just a teenie bit embarrassed for — and by — the tearful twitchers of Winspit. I mean, is that really a terribly, let's say, dignified image for birdwatching? It doesn't even sound like fun! Just frantic ticking. And that's another thing...we've all heard twitching compared to train-spotting. Well, no disrespect to ferro-equinologists, but I've always rather assumed most birders would regard that as a bit of an insult. But maybe I'm wrong. Maybe we're all brothers under our anoraks.

And maybe I'm sounding a bit too tetchy. Maybe it is old age. Maybe I'm turning into Dad, grumbling at the kids for playing their heavy metal albums. Call that music? Call that birdwatching?

Nevertheless, can I at least ask a few rhetorical questions? Why do twitchers twitch? Is it a competition against fellow twitchers? Who's got the longest list? If so, isn't it a bit meaningless? Except perhaps for the very top 10 of the 400 Club, which — let's face it — anyone can join if they're prepared to put in enough time, effort and money. I can understand grand obsessions — I'll personally cheer Ron Johns on to the 500 — but to the thousands in the second and third divisions I still put the question: why? Why the obsession with your British list? Local list I can understand. It's so much more personal. And I certainly sympathise with any Dorset dippers if they missed the Bluetail on their local patch. But I can't say I wept any tears for anyone who drove 300 miles on the day it pushed off. Anyway, they don't need my tears, they shed their own.

But is it really worth crying over? I dunno, maybe my feelings are getting numbed as the years roll by. Am I getting so weird that I can honestly say I got more pleasure last October from finding a Yellowhammer on Hampstead Heath that I would have done queuing up for the Bluetail in Dorset? Mind you, finding a Bluetail on the Heath...now that really would have been something else! An escape, in fact!

Then again, can we really be sure the 'Winspit Wanderer' hadn't just hopped out of a cage? Oh my lord. That'd be rather like discovering that Milli Vanilli didn't do their own vocals, wouldn't it? Reason enough to cry! No, come on, you'd have to laugh...wouldn't you?

PS. In case anyone construes this article as anti-twitching (or anti-Bluetails), lighten up! It's whatever turns you on.

Do the right thing
Put it on the pager...or not?

Sometimes you get it right. Early this October I decided 'to hell with work stress' and invited myself up to Norfolk to spend a couple of days with my mate Tony, who just happens to live in Cley. Yes, Tony is a birder.

The wind was from the east and had been for several days. 'It's been pretty good' said Tony. After a day at Holme and Titchwell, I reckoned it wasn't good: it was great. You may well feel a bit blase about Robins — surely they belong on Christmas cards and garden spades? — but believe me a 'fall' of them is really something. There must have been literally tens of thousands along the east coast during this period. Even local birders claimed they'd never seen anything like it, and for we inlanders it really was an extraordinary experience. Apart from anything else, these Continental redbreasts (not quite so red as the British version actually) far from being spade-perchers are skulky little blighters; which somehow makes them more exciting. It also makes them ultimately more exasperating, as they are capable of impersonating just about any rarity from a Flycatcher to a Bluetail. The woods positively echoed with birders cries of 'What's that?' 'Oh just another Robin'. The fact was though that rarities were not what this fall was about. As well as the Robins, there were Goldcrests literally flitting between our legs, and Chiffchaffs dangling under every sycamore leaf. Try sorting that lot into abietinus, tristis and so on. All I know is they didn't look British. And neither did the Song Thrushes. Their invasion had even affected my London patch where I'd recently seen 50 in a morning, having failed to find a single pair all summer. It surely makes you think. Song thrushes (and several other species) are undeniably in bad trouble in Britain, but there's obviously plenty of them wherever this lot were coming from. So where exactly is that? Presumably eastern Europe, where agriculture is not so intensive. Kind'a proves a point doesn't it?

The second day, 8 October, was wet, windy, but rather wonderful. We started at Salthouse, where the waves were threatening to breach the sea wall and the spray soon coated my spectacles. Nevertheless, a lens cloth wipe revealed the truly delightful image of up to 20 Shore Larks shuffling over the shingle, black and yellow masks literally glowing in the gloom. Somehow they managed to actually look warm. But then they probably were. After all, they're used to Arctic blizzards.

By the time we reached Stiffkey, so were we. Well it wasn't actually snowing, but it was blowing a gale and the rain was lashing down. 'You should have been here yesterday' was the inevitable greeting from the dozen or so bedraggled birders. 'Gorgeous Pallas's in that Hawthorn. Bright sun. Eye level. Hovering and showing the rump.' I pretended not to care, but the truth is that though I've seen lots of Pallas's in Britain, not one of them has ever exposed its little lemon posterior to me. Never mind, consolation was at hand. Or rather it was perched on a nearby caravan. 'Great Grey Shrike!' I yelled, just loud enough to scare it away before any of the Pallas's people could whip round and see it.

There were migrants in the sheltered spots, but the leaves were shivering almost as much as we were and birding was very difficult. By midday, we had to admit they'd been right: we should have been there yesterday. Or perhaps we ought to have been somewhere else at that very moment. Well, that's what Tony's pager kept telling us. 'Wryneck at Holme, Red Breasted Fly at Hunstanton, Yellow-browed at Wells etc etc.' I confess I have mixed emotions about pagers. In many ways I'd rather not know what I'm missing. (I'm not a twitcher, but I do have feelings!) But I have to concede that the efficiency of the service is quite incredible. What's more, we kept meeting people who were also definitely not what I'd call twitchers — in fact I might call them dudes — but who were nevertheless allowing their pagers to dictate their birding itineraries; and they were having a whale of a time, and seeing some very nice birds (that shameless bottom-flashing Pallas's for one). Nevertheless, my personal use for pager info is mainly to warn me where not to go. Plus there's the lure of negative information.

For example, there was no news from Blakeney Point, so obviously that's where we should have gone. That would have been the pioneering thing to do. I also suspected that it would have been the totally barmy thing to do, as sueda bashing in torrential rain and a howling gale is almost invariably an utterly futile and frustrating exercise. We decided to prove it by flailing through the narrow fringe of sueda that edges the saltmarsh at Stiffkey, rather than stroll along the path like everyone else was doing. In fact, it was rather fun. There were birds in there, but the views we got were almost subliminal. No sooner did they flit from under our feet than they were carried away on the wind, before plummeting back into cover. The glimpses were just about sufficient to identify the obvious: Robin, Dunnock, Goldcrests, Chiffchaff, Redstart — that was nice — and Siberian Stonechat! Eh? Yep, definitely. A lovely frosted first-winter, which even dared to perch for a moment. Then we remembered that it had been advertised on the pager. We just hadn't realised we were in the right place! It was nice to sort of find it ourselves.

So, the sueda hadn't been futile; but it was about to be frustrating. Even as I was admiring the Stonechat, something fluttered past my left knee. 'I've just flushed a streaky warbler' I called over to Mike, a friend of Tony's who'd joined us along the way. We walked back through the patch of sodden rank weeds into which the bird had disappeared. Nothing. Then something. 'That's it! Locustella. Streaky back. Big round tail. Whirry wings.' Gone again. Snippets of gen from a two-second flight view. Tony joined us and we tried again. Nothing. OK, one more go. 'There it is!' Up, down. Gone. 'What did you get?' Streaked Locustella. Looked big. Warmer rump.' 'I saw pale tail tip. Honestly.' The other two gazed at me. They knew what species we were all thinking of. What's more, I'd found one in Shetland many years ago. 'But', warned Tony cautiously, 'normal Grasshopper Warbler can show a pale tip.' 'And', I added authoritively, 'Pallas's Gropper's pale tips really aren't visible in the field.' But I still had a feeling this bird was one. So what was the right thing to do? We gathered the two or three soaking birders who hadn't already fled back to the shelter of their cars and walked the weeds again. And again. But nothing. We discussed putting it on the pager. No way, we decided. Things look so positive in print. Even if it says 'possible', people might fly in from Scilly or somewhere and lynch us if they don't see it, or — worse still — label us as stringers, if they do and it turns out not to be one. So we settled for calling Richard Millington at Birdline, and leaving it up to him and others to come and check it out.

Meanwhile, we simply erased the mystery bird from our memory bank (well, nearly) and headed for Wells Woods. The rain got heavier and the sky got darker, but not so dark as to obliterate a final unforgettable image of a great value day: as we were leaving, something made me look up and there, perched barely 30 feet away, was a Long-eared Owl, blinking down at us with those glorious orange eyes. Around us, 20 odd Robins clicked anxiously. In the distance, Pink-footed Geese were yelping; whilst Redwings teezed and Bramblings wheezed overhead. Pity the Yellow-browed didn't give us a tsooeet, but you can't have everything. Like that bloke on the Fast Show would say, isn't birding brilliant?

Oh, in case you're wondering, no one ever did refind 'our' warbler. Unless it was the controversial Locustella seen on Blakeney Point a few days later. But that's another story. And I'm not telling it.

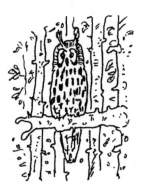

Value for money
The Wildfowl and Wetlands Trust

Saddest twitching tale of 1997 (and I don't mean the cruellest dip out). Welney in December, male Canvasback had just taken up residence. Twitchers travelled from far and wide. A £3 entrance fee was chargeable to non-WWT members, which allowed them as long as they needed to tick off the bird (about two seconds in many cases). The vast majority of them paid happily. But not one sad individual. Apparently, he sat in the information centre café clutching his pager in one hand and his money in the other, refusing to part with the latter until news came up: 'Canvasback showing in front of the hide NOW!' At which point, he handed over the cash and rushed across the road to see the bird. Would he have asked for a rebate if it had flown off, I wonder?

Actually, I'm grateful to this pathetic creature (the twitcher not the duck) for providing me with a starting point to this month's piece. A couple of points that the incident brings up in my mind. First of all, WWT: The Wildfowl and Wetlands Trust (the only downside to this splendid organisation is that it changed it name from the Wildfowl Trust to a tongue-twister rivalled only by the World Wide Fund For Nature; but I digress). First point: I'm glad most Canvasback fans coughed up their £3, but really they should have been WWT members in the first place. I've always maintained that any birder who is not in the RSPB should have their bins confiscated. I feel the same way about the WWT; and the BTO; and your local County Trust. I don't want to get preachy (though I am going to) but I really believe that anyone and everyone who gets pleasure from watching birds should join as many conservation groups as possible. (Well, those four will do for starters.) And please don't give me the old cost excuse. Come on folks, if you can afford a pager, a Birdline phone bill, top optics, and a subscription to *Birdwatch*, you can stretch to a few membership subs as well.

As regards WWT specifically: just think for a moment about what it has achieved over the years. Not just the conservation successes, the reintroduction programs, the scientific research etc etc. Not just the various sanctuaries and their very memorable birds — Slimbridge, Caelaverock, Martin Mere, Arundel, Welney etc etc. But also the basic principle of making birds accessible to people. It is arguable that of Sir Peter Scott's many legacies to the world and wildlife the one that people most consistently benefit from is his pioneering belief in providing facilities for us to actually see and enjoy the birds. Hides, scrapes, Walkways, bird reserves etc. WWT were — and still are — very much leaders. We should all be grateful. Own up, and pay up.

Second point. Back to Welney. I happened to spend a day filming their in mid-December. First of all, yes, we saw the Canvasback, and yes we got picture of it; though for all the action it provided it might as well have been stuffed. On the other hand, it looked a pretty genuine wild bird, in so far as it didn't come to be fed (though by the time you read this it might have taken to perching on visitors' shoulders and quacking 'who's a pretty boy?'). But, as it happens, it isn't actually that pretty. Not as pretty as a Pochard, for example.

Which brings me to the real value of Welney on that day. The weather was superb and the light stunning. And what a view you get from the centrally-heated, picture-windowed main hide. OK it's a bit dude I suppose, but on a raw winter's morning, don't knock it. Gaze and enjoy. And start with those Pochard. Hundreds of them, mainly males, glowing in the sunshine and packed together like an Escher lithograph. Beyond them, packs of Pintail — is there a more handsome duck in the world? — and great swirls of Lapwing, every now and again panicking as a Sparrowhawk zoomed across.

All very well, but a bit basic perhaps? OK then, you want a proper birder's challenge? How about searching through a few thousand Teal till you find the Green-winged? Or scrutinise a scopeful of tundra Bean Geese. Zoom in on the bill patterns, sort out the juveniles, take in the jizz, to compare with taiga Beans next time you see them. You can even spend more than two seconds on the Canvasback. Don't ask for directions, try and find it yourself. Not so easy if it's asleep, which it often is. Notice how the light makes a huge difference to how pale the body looks (ironically, it's much easier to pick out when the sun goes in). You might even care to point it out to some of the non-twitchy people in the hide.

And talking of them: don't be too proud to stick around for the afternoon feed (of the birds, not the people). I know it may seem

a bit like a zoo, but it's not. These are wild birds feeling safe —
and hungry — enough to accept human help. Maybe that's why
they called it the Wildfowl Trust — 'cos they do! And, be honest,
isn't it pretty amazing to be only a few yards away from
Whooper Swans? Though not quite as amazing as the Welney
finale, which I defy any birder not to be moved by. You need to
wait till it's almost dark, except hopefully for one of those 'Peter
Scott' sunsets in the west. That's when the Whoopers fly in to
roost on the washes, after feeding way out on the fields for most
of the day. It is quite simply a great sight. Then step outside the
hide and listen, and it's an even greater sound.

All that for three quid? Compared to what you'd pay to see
some crappy West End musical, or naff football match? Come
on, it's New Year's resolution time: if you are not a WWT
member, join now.

Birdlines

What do you want to hear?

I'm always blathering on about how I'm not a twitcher. So how come my itemised phone bill consists of an almost unbroken column of 0891 700 something-or-other numbers? Am I a liar, or a masochist? Or do I keep a telephone list: 'birds I've heard mentioned on the phone?' (I'd be a member of the 'Five million club' on that one.) No. But I do call Birdline an awful lot. For a variety of reasons. 700222 (National birdline, of course) I ring just so I can keep sort of 'up to date'; and in case anyone asks me 'what's happening in the bird world?' Such an enquiry rarely comes from birdwatchers (who obviously call birdline themselves) but from people who are either being sociable or silly. In the latter category comes blokes on building sites, and taxi drivers. 'Oi, Biw' (Biw is the matey version of Bill; say it out loud and you'll get the idea.) 'How's the old birdwatching going them, eh?' I can react in one of four ways. I can ignore them, and risk being called a 'miserable bleeder'; or admonish them for their pathetic sexist banter, and risk a punch on the nose; or I can smile, and risk them starting a conversation. Or I can tell them. 'Well, since you ask, apparently the female Bufflehead of unknown origin is still at Hornsey Mere, viewable from the RSPB hide at Kirkholm Point.' That usually shuts them up! Except on the occasion when a six-foot scaffolder snapped back at me: 'Yeah, but its got a metal ring on its right leg. Obvious escape.'

More often, though, I ring local birdlines. Particularly if my wife and I are off the visit Auntie Joan, who just happens to live in a very small cottage near East Dereham, which just happens to be within half an hour's drive of most of the birding hot spots of Norfolk. Heaven knows, I thoroughly enjoy watching six hours non-stop daytime telly in a room the size of a postage stamp and in a temperature unsurpassed outside a sauna or the Gobi desert at noon, but I am grateful that a call to 700245 (Birdline East Anglia) gives me an excuse for the occasional escape. 'Oh, sorry Auntie Joan, but there's a Pallas's at Cley. You don't mind if I nip out for a few hours do you?' She always smiles and says: 'Of course not. Off you go.' No doubt, she's only too glad to get rid of me; as indeed would be any auntie whose step-nephew's idea of a cheery visit is to pace up and down, sighing loudly, and muttering about 'claustrophobia city'. Nevertheless, I hope Auntie Joan doesn't read this, because the truth is I am deceitful even in my excuses. As it happens, my calls to 700245 are entirely negative. Having heard about the Pallas's, Cley will be the last place I nip off to. Neither will I go anywhere else that is hosting a current twitch. I call Birdline to avoid the crowds. Whichever place isn't mentioned, that's where I'll go. I call it enterprising; others call it perverse. It's definitely unsociable. Mmm, maybe I have a problem with people. Sorry, Auntie Joan.

Anyway, the Birdline I call most is 700240: Birdline South East. 'Cos that's where I live. In London, as it happens, and — as regular readers of this column will know (possibly to the point of irritation) — my local patch is Hampstead Heath. In fact, I rarely go anywhere else. So why do I call Birdline South East? To hear my own records? Well, partly — come on, who doesn't? 'In central London, a brown-head Goosander on Hampstead Pond'. That was mine, that was! (Well a Goosander's a mega-gripper for the Heath. That bad, huh?) But mainly I call so that I can put my observations into a context. (Uhuh, 'observations', that's a big word, must be serious ornithological bit coming up. Yep, you'd better believe it.) However, I have to say, complain almost, that I don't always hear the information I'm hoping for. I'll give you an example…

Flashback to Sunday 12 January. After about three weeks of frozen, and largely birdless, conditions, a sudden thaw set in. The effect over the Heath was immediate, and, in its small, local patchy, way, quite spectacular. Not only were the trees suddenly alive with the sound of singing tits and drumming woodpeckers, the skies were full of winter thrushes. Hundreds of Redwings and Fieldfares went streaming over, travelling from east to west. I assumed that this wasn't just happening over Hampstead, but I wanted confirmation. So I rang Birdline South East. And what did I hear? 'Of national interest, the female Bufflehead, of unknown origin, is still etc etc!' I screamed, but I didn't hang up. Eventually, we got to the local stuff. 'The Ring-necked Duck is still in Surrey docks (followed by a minute's worth of map references and access instructions)…the four Bean Geese are still at Theale (followed by another minute of less impressive species that were 'still at Theale')…the Bittern is still at Frensham Great Pond (followed by half a dozen more Bitterns that were still somewhere else). Plus, various other birds that were still wherever they had been for the past several days, or weeks, or possibly years. No mention of Redwings or Fieldfares. Or maybe there would have been, if I'd hung on for two or three hours, by which time my next phone bill would have necessitated an application for a millennium grant. Or maybe that's the point. (Oh, what a blind, blind fool I've been.) No wonder those Birdline chappies drive such posh cars. No, that's not fair. If they are doing well, they deserve to, because Birdlines (all of 'em) do a terrific job.

Nevertheless, I will finish with a little request. How about the first part of the message being a brief, very brief, if you like, summary about what's generally happening in the region? Like: 'Welcome to Birdline South East, Bugger all today. Stay in and watch the telly. Or why not give Auntie Joan a ring? It's good to talk.' Or, on a better day, and quite seriously, how about a rapid resumé of movement or migration? (And not just for the arriving spring migrants.) Or maybe there could be a separate number to keep us local-patchers up to speed. Or maybe there is, and I don't know what it is. But if not, why not? OK Birdliners, we know you must read *Birdwatch* — we certainly read *Birding World* — so how's about a response? A chance for a plug, and to make an already excellent service even better.

"Little oik, more like!"
Ticked off...at last

Do they still call them 'sprogs'? Relatively common birds that you still need for your British list. Elusive little blighters like Hoopoe, Golden Oriole or Lesser Spotted Woodpecker. For many years, mine was Little Auk.

As a juvenile, I had an excuse. I lived and birded almost exclusively in Birmingham. Not prime Little Auk habitat. Mind you, every winter the local paper published photos of wrecks of Little Auks frolicking in someone's bath after falling down a chimney or plummeting into Cannon Hill Park. In fact, all through my school days, I was convinced that one day a Little Auk would fall on me. But it didn't.

For the next 10 years or so I didn't panic. I was sure I'd pick them up on a seawatch some day. I was often in the right place at the right time. October in Norfolk, November in Shetland. I scoured the seas and I stared at the Starling flocks Little Auks are meant to tag along with. But no luck.

I got close. I missed them by a day, an hour, even a minute. I once sat with a friend on an Irish headland while he watched one fly past which I couldn't pick up in my scope. He reassured me: "They never come in singles. There's bound to be more." But there weren't. Not for another 10 years there weren't...

I was 46. Middle-aged crisis. I could no longer go on living a Little Auk-less life. On 9 November 1987 I made a phone call that made me realise just how desperate I'd become. For the first — and last — time I rang Birdline for other than a vicarious thrill or to find where to avoid the crowds. The voice of destiny (or was it Lee Evans?) summoned me: "90 Little Auks off Flamborough Head this morning."

As if to prove for once and for all that I'm not a proper twitcher, my conscience told me I couldn't just race off to Yorkshire. I needed an excuse. I got my agent to ring Yorkshire Television and book me on to the panel of Through the Keyhole. God, how low can birdwatching drag you!? I was due up there on November 11. Would the auks still be passing the Head?

On the morning of the 10th I rang Birdline and checked the weather forecast to reassure myself. The message was mixed: "291 Little Auks of Flamborough this morning". Good, but, "East wind becoming strong south-west". I hadn't yet seen a Little Auk but I knew their perverse little ways. They hated south-west winds. I was going to dip out…unless I could get up their very quickly indeed. During the next 24 hours I discovered the real reason I can never be a twitcher. I can't stand the stress.

At about 10.30am on November 10 the nightmare began. I raced through London traffic towards Kings Cross. Every traffic light seemed to be red, every road was up. Would the station car park be full? Worse. It was "out of order"! I drove straight through the red-and-white ribbon and just dumped the car. (As it turned out, I got three days' free parking!) I leapt on to the 11 o'clock train, which then tantalised me by not leaving till ten past. For the next two and a half hours I could do nothing except get indigestion eating a BR sandwich, while watching a murky day turn into something approaching fog.

At York, I picked up a hire car and raced towards Flamborough and the gathering gloom. I ran to the headland. A lone birder was leaving. "Have you come for the Black Guillemot?" he asked. The what? Don't tell me that was his sprog! I resisted snapping that I had seen a million tysties in Shetland and instead enquired, as if casually, whether there'd been any Little Auks lately. "Oh, about 100…this morning". He left. I stayed. Till dark. I saw two Goldeneye and a possible Grey Phalarope.

Night in a local hotel. Dawn, back on the headland, where I discovered two more Little Auk facts. They don't entirely hate south-west winds, and they do prefer early morning. Within an hour I saw about 30 of the little critters, belting past at vast distances and looking like incredibly boring waders. (They were Little Auks though. I'm sure of that because I had one of Flamborough's regular seawatchers sitting next to me to make sure I wasn't stringing.)

So ended my quest, and the last of the birds in the first part of the Shell Guide was ticked off.

But here's the funny thing about it. Whenever I close my eyes and try to 'see' those Little Auks flying by…I can't. Maybe that's another reason I'm not a twitcher. Birds I don't find for myself just don't 'stick'. Anyone else have that problem?

THE BEST AND THE WORST

This section needs no further explanation, read on...

Flamborough, October 1988
A cure for the wintertime blues

As I write, it is early January and I'm suffering from SAD. You know, that syndrome that afflicts people during the doldrums of mid-winter, when memories of autumn have faded and spring seems far away. I'm not sure what the letters stand for, but — having just rung Birdline and trudged round my local patch — it may well be Sod All Doing. At such times, I am always delighted to dive into the vicarious pleasure of a bird magazine. Today, I received the latest copy of *Irish Birdwatching*. I'm very fond of this one, for lots of reasons, not least because it is relatively unexplored. There's a freshness about the style, an excitement in discovery, a sense of real enthusiasm. It also features a regular 'My Favourite Day's Birding' item. A cliché, you might say, but you'd have to be jaded or jealous not to enjoy reading such accounts. For a start, it's fascinating to discover which elements different birders find particularly memorable. To some it's a huge 'fall', to others a successful twitch, to many the thrill of finding a good bird for yourself. Best of all, all three.

Which reminds me of...Flamborough, 17 Oct 1988. To be honest, everything pointed to it being good. The night before, I'd rung Birdline which told of an east coast riddled with rarities from Fair Isle to Sandwich Bay. But there was no mention of Flamborough. Which was precisely why I was going there. Before I went to bed, I made one last call to Andrew Lassey, the resident Flamborough 'guru'. In fact, Steve Rooke answered the phone. He was staying with Andrew for a day or two. Yes, they'd already had good stuff. Radde's, Pallas's and a record high of Yellow-broweds. Plus an even more remarkable low of birders. "There's about four of us," said Steve. It would soon be five!

On the 17th, I was up at 4.30, on the road before 5.00, across the Yorkshire border by 7.00 and, on the stroke of 8.30, I skidded into the car park at Flamborough's South Landing. Wind in the east, mist in the air, birds in the trees. Within half an hour, I was beaming at a binful of a Firecrest, two Yellow-broweds and a Pallas's flashing its rump more shamelessly than a stripper. That was the twitch part. Except that I was on my own. I drove off to find someone to share it with.

By Old Fall hedge, I found Steve chatting to Mike, another local birder. Andrew Lassey had had to go into work. He couldn't have been very happy about that, and we sympathised. We set off along the hedge, vowing to find something for him rather than to grip him off. There were birds everywhere. Redwings leapt around the hawthorns, 'continental' Robins flitted along the ditches, a Ring Ouzel chack-ed ahead of us, six Snipe zigzagged overhead, four Swallows hurried past, three Lapland Buntings ticked around with a pack of Bramblings, and two Sparrowhawks caused mayhem as they zoomed into Old Fall Plantation scattering a shower of Goldcrests and another couple of Yellow-broweds. Never had cliché been so apt: it was all happening. That was the fall bit. Now for the 'find your own rarities' experience.

Just after midday, Steve and I were scanning an enormous ploughed field. "I've got a bunting," announced Steve. "Where?" "Well...it's out in the middle. Half-hidden behind a sandy-coloured stone." Amazingly, I chose the right stone out of about a thousand! All we could see at this point was a stripey mantle. "It's probably a Reed," muttered Steve, with admirable caution. Then it popped its head up. A mass of markings. "No it's not." "Is it a Lap then?" It shuffled out from behind its stone and turned to face us. We spoke as one: "It's a Pine Bunting!" At which point, it flew. OK, we both felt confident of the ID but we'd seen it for barely a few seconds. No way could we write a convincing Rarities Committee description. ("Why was it a Pine Bunting?" "Because it was.") And it was still flying. Away over the hedge. It must have heard our sighs.

It turned, came back, and plonked down in front of our waiting scopes. "OK, so now have you seen enough?" And away it flew again. Never to return.

Steve raced off to phone Andrew and give him news that probably made his day worse. Meanwhile, a Great Grey Shrike provided me with 'interval' entertainment, by catching and devouring a vole before my very eyes. Then Steve returned, and we were off again. Along another hedge, across to another field. This one was being ploughed by no less than three tractors. "The farmer's getting it ready for a rare wheatear," I suggested. "Like that one." It was worth a try. The power of positive thinking. But the Wheatear was only a Common (sorry, Northern). So I tried again. "How about that one then?" A second Wheatear hopped out of a furrow. This time — believe this — it was a young male Pied. Off raced Steve to call Andrew, while the Pied lined itself up on a fence, alongside its common cousin and half a dozen Black Redstarts, including a stunning male with dazzling white wings. And the sight was enjoyed by all, including Andrew.

So that was it. Good birds, good deeds, good company. And I feel a lot better. Maybe that's what SAD stands for: Simply Amazing Day.

Grow your own!
Lundy, October 1984

OK. I admit it: I love finding rare birds. Chasing other people's doesn't do much for me, but I confess that, in my eternal quest to 'find my own', I am undeniably as frantic as the most rabid twitcher. The secret of finding your own rarities is, of course, to go to places where other birdwatchers don't. There is an obvious flaw in this strategy. There is usually a very good reason why other birdwatchers don't go to these places: like, they are not very good for birds (let alone rare ones). However, another reason people may not go there is that some of them might be rather inaccessible, and true twitchers can't stand the idea of being 'stuck' somewhere, so they can't instantly race off and follow the call of their pagers. This is why I love small remote islands. A third thing I love is...American warblers. All of which adds up to a long cherished ambition — no, obsession — to find my very own American warbler, on a small island, which wouldn't instantly be invaded by other birdwatchers.

Which brings me to Lundy, October 1984. I flew in on the 6th. Yes, I said 'flew'. This was back in the wonderful days when a little helicopter — the one used by Anneka Rice in Treasure Hunt, but without Anneka in it (you can't have everything) — whisked you across from North Devon in 7.5 minutes. Oh joy! Compare that to the couple of hours gut-wrenching agony on the SS Oldenberg which you have to endure nowadays. As I arrived, a party of ringers was departing. 'Not a lot of birds', they told me. Such was my unshakeable optimism that I actually recorded this in my first night log book as a 'plus'. 'Nothing to chase. Relieves the pressure. Now I can find my own birds.' There's positive thinking for you!

The very next morning, I was rewarded. Wandering around in the gloom of first light I heard a soft 'tick' go over my head. An hour later, with the sun up, it flew across again and pitched down on a thistle, revealing itself to be exactly what I'd suspected: a Little Bunting. Later in the day, I bumped into the only other birder who was staying on the island. He called me a 'jammy bugger!' Fair enough.

Actually, there were a few more birders on the island, but they were ringers, so they didn't count. No disrespect intended, of course. All I mean is that the 'Lundy ringers' (who usually come from somewhere up north I think) rarely stray outside the little sheltered valley of Millcombe, which is entirely surrounded by mist nets. Come to think of it, maybe the ringers can't get out. Fortunately, birds can get in, and, if any of them end up in the nets, you are guaranteed a close up in-the-hand view. (Though that doesn't always mean the bird gets correctly identified. Oops! Naughty reference to embarrassing Thrush Nightingale/Veery incident there. Sorry). Anyway, call me an old purist if you like, but I find ticking off in-the-hand birds about as valid as 'counting' the stuffed ones in the Natural History Museum. What's more, there's a lot more to Lundy than Millcombe, lovely spot though it be.

So, having relocated 'my' Little Bunting on a ten-foot-square ploughed field right in front of our cottage, I left the ringers to surround it, whilst I set off to do the rest of the island. That day, I didn't find much and, to be honest, the highlight was to get back and see...the Little Bunting in the hand! However, the next day was magical.

It was sunny and warm and absolutely dead calm. I think the Met men call these conditions 'cyclonic'. I've also learnt to call them 'promising'. So it immediately proved. As I left the house, a Chiffchaff flitted out to greet me, followed by Goldcrests, Meadow Pipits, and alba Wags. It wasn't a deluge, but it was definitely a fall. It was also a rather odd selection. One of the commonest species was Stonechat. Everywhere I scanned, there were little parties of them, up to 15 in a group, flicking around the tops of the bracken. I estimated at least a hundred on the island. Meanwhile, the ringers were busy bagging bunches of Blackcaps in Millcombe. I left them to it, and set off to check the other sheltered valley that serrate the east side of Lundy. My favourite is called St Helen's Combe. I sat at the top of the slope, looking down on a large sycamore. The leaves were motionless, but still managed to hide a number of 'crests, and several species of warbler. But they were all British. Or were they? Suddenly, I caught a glimpse of something 'streaky, with a couple of wing bars.' A Siskin? No way. Too big. Too elusive. I sometimes think it should be one of the diagnostic features of some rarities: 'If it keeps disappearing, it must be rare.' It took me an hour and a half before the bird finally gave itself up, and sat out in full view, preening and allowing me to check all the tricky little details of a 'confusing fall warbler'. It was a Blackpoll. Not perhaps the rarest of the 'Yanks', but it was mine, all mine. I loved it.

I loved it even more a few days later. I was driving back to London, when a car load of birding friends passed me and waved me down into a lay-by. They'd been on Scilly for a fortnight. 'So what did you get?' I asked. 'Oh, we had several Yanks — Semi P. Solitary Sand, Swainson's Thrush, Yellow-throat, and (yes) a Blackpoll — but...' 'But what?' They really didn't seem very happy. 'It was all 'old stuff'. Most of it's still there, if you wanted to go down and...' 'No thanks,' I replied. 'I've just had a week on Lundy. Little Bunting, masses of Stonechats, and a Blackpoll. It was great.'

I really do think they envied me!

Three of a kind

A hat-trick of Sprossers...in one week

In early September I went twitching. Well, sort of. In fact, I had to go to Suffolk to do a wildlife promotional morning at Alton Water (a splendid and seemingly neglected reservoir, by the way). Alton just happens to be within nipping-in distance of Felixstowe, so I and Derek Moore (then Director of the Suffolk Wildlife Trust) thought we might as well go and pay our respects to the Thrush Nightingale that had been in residence at Landguard for a week or two.

I wouldn't have called it a 'crippler', but at least it knew how to behave. It has always struck me as doubly irritating that most really skulky rarities are also incredibly boring to look at. Maybe they are embarrassed to show themselves. However, the Landguard 'Sprosser' showed no such coyness, hopping around as bold as a Robin and so faithful to one small bramble patch that one could have accused it of having an appallingly low sense of adventure. I suspect it may have ended up dying in there. But not before it had given many a birder a useful tick. But not me. I'd already seen three British Thrush Nightingales. All at the same place and during the same week. And what a week it was.

Flash back to Fair Isle, May 1970. On 5th an anticyclone settled over Scandinavia. For a day or two the south-east winds blew but nothing happened. Then, on 7th, it started. Willow Warblers and Tree Pipits rose from single-figure counts on 6th to 100 of each on 7th. With them came the indicators that something special was happening. A dozen Wrynecks. A dazzling Black-headed Wagtail. In fact, I didn't catch up with that one till the next day because at the time it was found I was enjoying my own 'magic moment'. In pioneering mood I had marched off to the north of the island, while everyone else went to the south. As the sun began to set, I'd seen very little and was beginning to think I'd taken the wrong option. Weary and rather disappointed, I slumped down on the clifftop near the North

Light and started nibbling a consoling Mars bar. As I gazed down at the beach below, my eyes began to focus. It was as if the cliffs were coming to life. A Redstart flicked down onto the seaweed. Every damp patch had a Willow Warbler on it. Clearly birds were arriving, 'hitting' the cliff, and slowly working their way upwards. Suddenly, hardly a yard away from me, two Bluethroats popped up over the crest, their breasts looking for all the world like Union Jacks they were about to plant at the summit. I almost cheered. I had a feeling that the next day was going to be pretty amazing.

It was. Not least because the weather was delightfully clear and sunny. On Fair Isle itself, at any rate. However, some miles to the south there was a nasty patch of fog and drizzle. The result was that on the island it rained birds. What's more, the deluge continued for two days. The figures conjure up some of the picture. Wheatears: 2,000 on 8th, rising to 4,000 on 9th. Redstarts: 300 on 8th, 700 on 9th. Whinchats: 350. At least 1,000 Willow Warblers, 500 Tree Pipits, 35 Wrynecks, 14 Red-backed Shrikes, 10 Ortolans, and no less than 35 Bluethroats.

But, in truth, it isn't the statistics that really stick in the memory so much as certain images and incidents. I shall never forget looking up as a cascade of Tree Pipits plummeted down so abruptly that I instinctively ducked. Or tiptoeing across the green sward at the top of the West Cliffs having to take great care not to tread on weary Willow Warblers. And the 'weirdos': a single Corncrake zooming past me like a demented chicken, and a Great Spotted Woodpecker frantically looking for non-existent trees.

And, of course, there were the rarities. As is so often the case, many of them were found the day after the big fall, when the distracting curtain of commoner birds had lifted. A Short-toed Lark on 10th, and — quite incredibly — both Spotted and Little Crakes, not only on the same day — 11th — but even in the same ditch!

And what about the Thrush Nightingales? One a day, three days running, but definitely different individuals. The first was, typically enough, lurking under a rhubarb patch trying not to be seen. The second I found myself, on the grass in a sheltered 'geo' (a precipitous coastal hidey-hole), relaxing in the sunshine: both of us, that is, me and the bird!

And the third I actually caught. Well, sort of. It had spent the morning leading the observatory staff a merry dance round a large field, which was gradually becoming surrounded on all sides by walls of mist nets. However, even as the Warden had his back turned tying off the last pole, the bird made a bid for freedom, zooming past him and diving into the safety of some derelict farm buildings. Or so it thought. It landed on the doorstep of an old barn, heaved a sigh of relief, and turned back to laugh at its would-be captors…at which point I ran 50 yards straight at it, waving my arms and 'pishing' wildly. Understandably, the bird turned tail and flew into the barn, where it took a sharp left and flattened itself against a window, thus temporarily knocking itself out! Happily, it soon recovered and was sent on its way bearing a shiny new ring.

I'd like to think it is alive and well somewhere even now, but that's unlikely, since it would be at least 25 years old. Nevertheless, perhaps it passed on the benefit of its experience to later generations of Thrush Nightingales: "Listen, if you don't wanna get hurt…either stay out of sight completely, or perform for the birders." That Landguard bird had learnt the lesson well.

Strangers in the night
Fireworks of a different kind

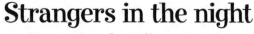

I am writing this on 6 November — the day after Bonfire Night. We had rather a good 'do' this year at a street party involving so many people making a small contribution that the fireworks display would have done credit to Jean-Michel Jarre. Plus, there were baked potatoes, parkin and garlic bread (well, it was in Hampstead), and only one child got slightly singed. As I strolled home, I savoured the traditional noises of the night: bangers, rockets, laughter, screams, fire engines. The only thing missing was the sound of birds. But why should I expect to hear birds on Bonfire Night? Well, it's all to do with one of those never-to-be-forgotten formative experiences from long ago.

Cue flashback music. Screen goes wobbly. Pages of calendar fly by, as the years regress to 1962. The Beatles are climbing the charts, and I am wandering the streets of Cambridge. Yes, dear reader, hard to believe it perhaps, but I did actually attend the light blue varsity. What a waste of an education, eh? Actually, I was rather active during my three years — I played rugby, jazz and the fool, and I even got a degree — but I never did feel I really 'belonged'. Consequently, I was often lonely. Hence the street wandering. I was particularly prone to this on supposedly festive nights, when I would amble round town punishing myself with the thought that everyone else was having a really good time and fantasising about being invited to join in.

Bonfire Night 1962 was proving a particularly bleak experience. By 11 o'clock I had given up all hope of sociability and was heading back to my room to indulge myself in playing Wagner, reading Keats and smoking opium...or whatever self-pitying students did in the early Sixties — Miles Davis, Health and Efficiency and a bottle of brown ale, probably. But even as

I trudged across the lawn of New Court, Pembroke College, my melancholy was dissolved by joyful noises. (Just 'cos I'm an MA Cantab Eng Lit doesn't mean I can't mix my metaphors.) The sound of distant fireworks was being drowned by bird calls.

OK, Redwings migrating at night is hardly a one-off phenomenon, but I'd never heard anything on this scale before. What's more, it wasn't just Redwings. Fieldfares, yes, I'd expect them. Blackbirds and Song Thrushes, fair enough. But Bramblings and Skylarks and a single Snow Bunting: do they normally travel at night? Well they were definitely up there, chirruping, squeaking and tinkling away. I was totally entranced. For two hours I stood rooted to the spot, attempting to count calls and identify species.

The movement ebbed and flowed. From 11 to midnight it was very lively, then they took a sort of half-time break, before another surge from 12.15 to one o'clock, at which time the skies went suddenly silent. At peak times, I was hearing at least 100 Redwing calls a minute. Wall-to-wall Redwings, really. Fieldfares were nearer 20 a minute. The other species were much more sporadic. As indeed were the waders. Oh, I haven't mentioned them yet, have I? Curlew, Lapwing, Snipe, Golden Plover and a surprisingly large number of Dunlin. I wrote in my notes: "Dunlin. Heard at least 25 times. Several obviously singles, but at least 50 per cent sounded like small parties, and about four or five sounded like large-sized flocks. Especially from about 12.15–12.20 the sky was full of Dunlin calls."

So why were so many birds on the move, and where were they going to and from? Well, in fact, the weather conditions were pretty classic. A whacking great 'high' over Scandinavia, strong east winds all day, a bit of rain, then a calm night. And, of course, as the Redwing flies, Cambridge isn't a very long way from the recognised arrival points of The Wash and the coast of north Norfolk. Nevertheless, think what scale that movement must have been on. I was standing in the quadrangle of a college bang in the middle of the town, surrounded by miles and miles of flat land. Not exactly an obvious hot-spot for migration. Presumably, birds must have been flooding in on a broad front over the whole of East Anglia. So how many were there? I wonder if anyone else was lonely enough that Bonfire Night to have been out there trying to count them.

Alas, in Hampstead on 5 November 1995 there wasn't a squeak to be heard. Maybe I'll never hear anything like it again.

Unless Standard cares to take up my challenge. How about a new range of fireworks? 'Cascade of Redwings', 'Fountain of Fieldfares', 'Deluge of Dunlins'. I'll buy a boxful.

Fairytale birding
Holy Island, October 1994

Last October I endured a particularly novel form of torture. I spent the second weekend of the month at the Northumberland Bird Fair. No, that wasn't the torture — that was as enjoyable as bird fairs always are. However, while I was 'stuck' at Druridge Bay the Northumberland Ringing Group was gallivanting around Holy Island on their annual get-together.

Holy Island is actually only about 30 miles north of Druridge, so you'd assume that if there were good birds there, there would also be good birds at or around the fair. But you'd be wrong. A freak front meant that Holy Island got the fall, but we didn't. In between chatting with mayors, signing autographs and trying out binocular straps, I kept sneaking out of the marquees to read the latest news from the island, which was being chalked up on a big blackboard courtesy of Birdline North East.

It was agony. Ten o'clock: "A large fall of thrushes and other common migrants." Then at regular intervals came the rarities: "Barred Warbler, Great Grey Shrike, Shorelark, Dusky Warbler." Eventually, I could take no more. I hid the chalk. I also vowed that the following year I'd be on Holy Island for the second weekend of October.

So this year it was all worked out. There was no bird fair at Druridge so I didn't have to feel guilty about copping out of it — and I had received an official invitation from the ringing group. Unfortunately, they forgot to invite the migrants, common or otherwise. The skies were clear, the wind was westerly and the weekend involved more bonhomie and booze than birds.

There was also a certain amount of Bill baiting. "Oh, you should have been here last year," "that's the bush where we had the Dusky," and so on. Frankly, I think it was very nice of me to stay and give my slide show on the Saturday evening but, then again, they were a very nice bunch of people. Oh, and by the way, there wasn't a mist net in sight — the Northumberland ringers' weekend is strictly birdwatching. Well, it would have been had there been any birds.

The ornithological highlight was arguably a Treecreeper (a bit of a local rarity), which was at least crawling along a wall so I suppose technically it was a wall creeper. Anyway, come Sunday afternoon and, after lunch and farewells, the ringers departed across the causeway, the tide came in and I was left gloriously marooned with Holy Island all to myself. That evening I tuned in to the weather forecast. Of course, what you really hope for is a brisk easterly and some drizzle or fog, with a consequent major avalanche of Redwings, Fieldfares and attendant more exotic fare. It's one of the ironies of birding: you go to a lovely place like Holy Island and pray for rotten weather!

In fact, what I got was the promise of 'a light south-east wind and some early mist patches'. Enough to raise my hopes. Monday would be my big chance.

I awoke to a day that was ominously calm, clear and — worst of all — quiet. Not a squeak or a chack greeted me as I wandered along the shore at 7am. I was instantly disconsolate. However, within 10 minutes I found myself enthralled by one of the most magical mornings I've ever seen. But hold it, twitchers, and fret not Northumberland ringers, this isn't the big grip-off. This has nothing to do with rarities. It has everything to do with ambience, atmosphere, call it what you will.

It also has a lot to do with Lindisfarne Castle. The guide book says that it was built in the 16th century by Sir Somebody-or-other, but then guide books always say that. The truth is that Walt Disney put it there. It is clearly the same castle that Tinkerbell flew out of. Or is it the one in Beauty and the Beast? Anyway, it's ever so impressive. The lowering turrets stand right on the edge of the ocean. Its only background is the sky. And that morning the sky was on fire.

As I approached from the west, the sun rose up through the mist behind the huge looming silhouette of the castle. I'm pretty derisive of that silly and overused word 'awesome', but this deserved it. OK, so Disney had provided the set; now Speilberg took over the direction. For the next glorious minute nature was stage-managed. First came a swirl of Jackdaws, cascading round a tower like bats. Then above them, a pack of Starlings. And then, bang on cue, a massive female Peregrine sliced through the lot of them. I almost applauded. Thoughts of falls of thrushes were blasted from my mind.

Then, as everything scattered, my eyes focused on a tiny dot perched on the very peak of the battlements. A single Redwing. What a brilliant touch of irony. At that moment I thought: "Maybe God has got a sense of humour". Or was it Stephen Speilberg?

"So what's your favourite bird?"

Shreveport, Louisiana, spring 1964

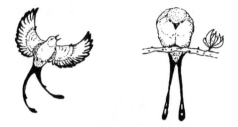

"And finally, Bill, tell us what's your favourite bird?" Every time I do an interview on radio they end up asking me that. Why? It's surely not 'cos they really want to know. I suppose it's what interviewers are taught at interview school: "Whenever you are interviewing a so-called expert, end by asking what is his or her favourite whatever it is they are an expert in."

Presumably they ask David Bellamy: "What's your favourite flower?" Or Patrick Moore: "What's your favourite constellation?" What do they do? Trot out some obscure Latin name and blind 'em with science? I must try that.

The thing is, if I answer the question, chances are that the interviewer — and most of the listeners — won't have a clue what I'm on about, unless I give them the names of something really familiar, like a Blue Tit or a Peacock. But if I did that I'd be lying. Frankly, Blue Tits don't do much for me — except recruit new members for the RSPB — and I positively hate Peacocks 'cos of that horrible noise they make.

So, when I get asked the 'favourite bird' question I usually come out with something 'smart Alec' or non-specific like: "It's the rarity I've just found". Or occasionally, I do offer a specific species and the interviewer says: "I've never heard of it," and then I try to describe it ,and probably everyone falls asleep, including me. After all, how do you describe a favourite bird in words? It's pretty difficult on radio, and may well be impossible on paper…but I'll try.

So what is my favourite bird? Well, I'm not being evasive if I say that I don't so much have favourite species as favourite images or memories. Some of them go way back, and they don't have to involve rare or even attractive birds. I'll never forget the winter-plumage Red-throated Diver that flew over my head last October — when I was standing on top of Hampstead Heath! Nor the female Golden Oriole (yes, female) dangling on a sprig of thistles on a spring evening in Cyprus. It was right out in the open, two feet above the ground, 30 feet away. You just don't see orioles like that. And all those canny colours: pink bill, lemony wash on the flanks, the purple flowers. Exquisite.

But if it's truly, mind-blowingly 'oh boy, look at that!' exquisite I'm after, I have to go back 30 years. In 1964 I was touring America with the road-show version of That Was the Week That Was. Those that remember TW3 on the telly are no doubt feeling their age at this very moment. Younger readers, be grateful for your youth, and don't worry if you've never heard of it. Suffice to say that the company I was travelling the States with included several 'funny folk' who are still going strong — David Frost, Tim Brooke Taylor...and me.

Naturally, from my point of view, the purpose of the tour was not to bring British satire to the Yanks, but to swell my American bird list. This I certainly did. By night I performed — in front of everything from 200 miners in Chicago to 6,000 students in Texas. By day, I birded where we happened to be. As it turned out, the itinerary was disconcertingly illogical, involving criss-crossing America in anything but a straight line. Thus, my bird sightings were equally disorientating. I recall gasping at Snowy Owls — in real snow, somewhere in Canada — one morning, and marvelling at hummingbirds in Florida the next afternoon. The net result of three months of this sort of thing was that I still have a much longer American list than British. I also have a veritable kaleidoscope of memorable images. I have only to close my eyes to 'see' again that flotilla of Surf Scoters bobbing around Vancouver sea front. Or the Mountain Bluebirds flitting along the telegraph wires near the Grand Canyon. Or the Vesper Sparrow on a gate post, or the Painted Bunting on a rubbish tip. I have no idea where those last two were.

But I do remember Louisiana. I recall the southern hospitality. "Y'all welcome here," beamed the taxi driver. And I'll never forget the chillingly instinctive racism. "Sure is pretty countryside. Pity it's spoiled by the niggers picknickin". We did one show, and it went great, with an entirely white audience, though we didn't have the nerve to include the Ku Klux Klan sketch! Frankly, I couldn't wait to get back to the airport. When I did, there was reward. Round the tiny terminal at Shreveport, there was a grove of magnolia trees. Every one was in blossom. The scent hung on the air, and the flowers delighted the eye — great generous bunches of them, gleaming white against a clear blue southern sky. And among the trees was the bird that was instantly to become my favourite: Scissor-tailed Flycatcher.

Have you ever seen one? Have you at least seen a picture? Take a look in the American field guide, but before that, let me try and describe it. The body is about the size of a Song Thrush, but the tail feathers are two or three times as long again. But they're not 'just' feathers — and they're not scissors either! They're streamers, or ribbons maybe. And the colours: subtlety incarnate. Soft dove-grey above shading to creamy-white below. And the pièce de resistance — which no still picture can capture — a pink flush on the breast, which deepens under the armpits so that, when the bird flies, it flashes startlingly scarlet, But Scissor-tailed Flycatchers don't fly — they twirl, they tumble, they dance. That morning there wasn't just one — there were over 50. Set them among magnolias and you've got a vision of heaven. Somehow, it's a bit lost on the radio!

Forgettable...
...that's what they are

One afternoon, late last October, I was wandering through Wells Woods. The only thing I was enjoying was the solitude. It was mid-week and pouring down. I saw no other birders.

I did see a flock of Redpolls. Most of the time they trilled over my head at great heights, as if defying me to get a decent view. Then, suddenly, they got bored with that game and plummeted down into a relatively small birch tree. I was able to stand directly underneath them and scan through the flock. It was clearly a 'mixed bag': juveniles and adults, and including several 'Mealies'.

"I wonder if there's any Arctics up there," I mused, even as I focused on the next set of undertail coverts. They were pure white. I was so close I could even get my scope on them. Not a streak in sight, except a few faint ones down the flanks. What of the rest of the bird? Well, grey rather than brown, a couple of decent wing bars, maybe a tinge of ginger on the face. Presumably a juvenile.

It was a close view, but not a great one: the light was lousy, the rain was on my lenses, and the bird refused to dangle down and flash its rump at me. Nevertheless, I thought: unstreaked undertail coverts are meant to be diagnostic of Arctic, aren't they?"

Rather than reply, the bird flew off with its chums and was never seen again. Neither by me nor — I presume — by any other birders, since there were no 'official' reports of Arctic Redpoll from Wells last autumn. I didn't send in my record either. Instead, I assigned it to a new category: 'Birds not worth seeing.'

Actually, I didn't mean to be disparaging to Arctic Redpolls. I'm sure a stonking great adult Greenland 'snowball' is a dazzling sight, but I'm not so convinced about the ones that need magnifying glasses, tape measures and six pages of colour photographs to sort them out.

However, I was rather taken by this new category. I mentioned it to a birding mate. He immediately came up with a nomination: "Brown Towhee! Definitely not worth seeing."

Then he showed me a slide of a bird so meek and pathetic that I almost burst into tears. I hadn't actually meant 'birds so boring they're not worth seeing', but I couldn't resist suggesting a few candidates myself. Darwin's Finches: scientifically riveting they may be, but — be honest — most of them look like female House Sparrows that have just been put through the mangle. Various Australian crows, several babblers and most cisticolas: hardly worth traipsing across the globe for. They sound OK, mind you, but hardly lookers. Winter-plumaged Crested Coot: pointless to look at, and it sounds crap too.

But the winner has to be Brown-throated Sand Martin. Even the field guide insults it: "Very small. Weak flight. Dingy and insignificant, with no prominent features. Resident." I should bloody well think so. I don't suppose it dares poke its beak out of the tunnel for fear of being mocked by flocks of jeering twitchers chanting "Boring!" But, of course, it would still be a tick. And, as it happens, I still 'need' B-t Sand Martin myself. So how can I possibly relegate it to 'birds not worth seeing'?

Well, actually that brings me to what I really meant. Perhaps I should rephrase it as 'birds not worth ticking'. No such thing, I hear you cry. OK, then, 'Species we're not really sure about'. Ah ha! Now the ears are burning. Oh lord, I think I must be referring to this 'lumping and splitting' business. I swore I wouldn't get involved, but I do love that phrase. It sounds so cross. "He stomped out of the room, lumping and splitting as he went!" And, of course, it makes the list-keepers cross, too. Nevertheless, think about it: is it a truly satisfying experience ticking off a juvenile Arctic Redpoll? Or three species of crossbill!? Come on, let's own up, we all know there are no more species of Crossbill than there are Herring or Iceland Gulls. Kumlien's schmumliens, I say.

What the dickens is a 'good species', anyway? Do you honestly understand it? I used to think it was birds that didn't interbreed in the wild. But, as we all know, wildfowl will hump anything that quacks, so where does that leave us? With one big superspecies called 'duck'? I confess, I'm confused. Just as I'm confused about various supposed races and subspecies. Why do I see Stonechats with white rumps and frosty scapulars from Ireland to Spain when, if a bird turns up in Norfolk looking like that, it gets called 'Siberian'? And why, if there are three crossbills, aren't there 23 chiffchaffs? Well, the experts are working on that one, aren't they?

Meanwhile, I'm going to listen to my eyes. If I see another juvenile Arctic Redpoll I shall shrug and walk away. If I see an adult 'fluffy snowball' I shall stay and enjoy it. And I'll tick it, too!

Simply the worst
A birding holiday in hell

Birdwatchers are always writing about their 'best days' or most 'memorable holidays'. But what about the worst? I don't mean just a day's dip out. I'm talking about a really rotten week, probably made all the more tragic because you'd been looking forward to it for months. No point in blanking out these memories. Let it all hang out, I say. So here I go.

Time and place: mid-September 1993. Kilbaha, County Clare, Ireland. My fellow sufferers: Tony from Grafham (a seeker-out of under-watched places, patient and pioneering) and Owen from Sussex (a local patcher but — let's face it — also a certified twitcher). The theory: Kilbaha is actually a perfect place to spend an observatory-style birding week. There are gardens around the village to attract passerines; there's Loop Head up the road for visible migration; and — best of all — it's barely a mile to Bridges of Ross, arguably the best seawatching point in Ireland. A good chance of great birds. As long as the wind is from the north-west. Which it often is in mid-September. But not in 1993. OK, so here's my doleful diary.

Monday evening: I nearly miss the plane due to torrential rain and flooding on the road to Heathrow. Forced to dump the car in the 'business' car park. It'll be expensive. Hope it's worth it. Met by Tony and Owen at Shannon Airport. They've been at Kilbaha two days already. They talk more about 'the accommodation' than the birds. They keep giggling. When I arrive I see why. Kilbaha House isn't exactly a National Trust cottage. It's taken them two days to discover how to turn on the water and the heating, and to excavate the kitchen from underneath a ton of mouldy crockery. But what do you want for two pounds a day?

Tuesday: dawn at the Bridges of Ross. We need a cloudy day with a strong north-westerly. In fact, it's sunny, with a light easterly. There's very little at sea. Loop Head is more promising, with a few hundred Meadow Pipits giving at least an impression of visible migration, and four Lapland Buntings ticking overhead. The cottage gardens produce only a single Goldcrest, but a nearby small pool with a muddy edge has Curlew Sands, Little Stints and a Ruff. This is clearly a spot to keep an eye on for a Yankee wader (mid-September in the west of Ireland should get at least Buff-breasted Sandpiper). So far, I'm perfectly cheery, flushed with first-day optimism, but I can sense that Owen in particular is getting a little restless for a change of scenery and a few more birds. I suggest an afternoon excursion south across the Shannon to County Kerry, where we can visit several famous rare bird sites. We miss the ferry by seconds and have to 'waste' an hour waiting for the next one but, by about four o'clock, we arrive at Akeragh Lough only to find...it has gone! There is a sign saying "Bird Reserve", a patch of dried-up mud, and a man walking a dog. But no water. And no birds. One of the most famous Yankee wader spots in Europe has completely disappeared! (I still don't know what's happened to it, and if any Irish readers can tell me where Akeragh has gone to, please let me know or, better still, put it back!).

We then race down to the estuary at Blennerville (another hot-spot). The good news is that the tide is just beginning to drop and waders are returning from their roosts. The bad news is that our lost hour has caught up with us. Darkness falls. Ah well, never mind. We'd picked up a couple of Med Gulls along the beach and the day list isn't bad at all. We look forward to...

Wednesday: ...we shouldn't have done. Weather: clear and calm, again. Very little at the Bridges. The gardens produce a Willow Warbler; a 10-mile tramp round the Head produces a Chiffchaff — and blistered feet; the little wader pool produces nothing, a particularly strong high tide having drowned the mud and probably the waders.

Thursday: stunning blue skies and dead calm. Completely birdless. The local shopkeeper shouts to us: "Well, you're certainly lucky with the weather." We have a choice between strangling her or going to sleep for the afternoon. I choose the latter. Can't speak for Tony or Owen.

Friday: possibly the most frustrating birding day I've ever had. Skies totally clear. Wind: gentle southerly. Even the few pipits and Swallows left in the area are tazzing out of the country. We decide to try and do the same. We call Aer Lingus, only to be informed that we'll have to pay the full return fare again if we want to escape early! Owen pleads that he's suffered a 'personal tragedy'. Which is true. The airline is unmoved. We have to accept that we must make the most of a bad job, but how bad is it going to become!?

We soon find out. We set off on the Kerry trip again. Heat haze and miles of low tide mud along the Shannon make wader-watching impossible. Akeragh Lough hasn't reappeared. But at least we've timed it so that we reach Blennerville on the rising tide. Wrong. The tide is actually in freak flood, cutting off roads and all hope of sorting through the birds. On our way back to Clare we call into a pub to try and cheer ourselves up with a little local Irish 'craic'. The advertised ceilidh turns out to be a naff country and western band. A fitting end to a truly dire day. But at least the weather has been lovely. Until...

Saturday: howling southerly and torrential rain. In the morning we write notes (that takes about a minute) and leave the cottage 'the way we'd found it'!

By mid-afternoon we are approaching Shannon Airport. Maybe there's going to be a happy ending, or at least a little consolation. Shannon Lagoon (right by the airport) is an excellent bird reserve, good for waders and wildfowl. We have half an hour before the flight goes, and it's stopped raining. Or has it? Even as we pass through the perimeter gates, the heavens open in a deluge that would have had Noah racing for his hammer and nails. We can't even scan from the car without steaming up or getting soaked. We finally get the message, so we go home.

Saturday evening: My 'business' parking costs nearly 100 quid — I could have had a week at Cape May! Moreover, National Birdline tells of the English east coast almost sinking under the weight of rare migrants: "Best September fall for years." And Owen rings Birdline South East to hear the first bird up — on his local patch — is a Buff-breasted Sandpiper. I've got a feeling he won't be going back to Kilbaha in a hurry. And neither will Tony or I. Well, not until this September, perhaps. After all, it couldn't happen two years running...could it?

RINGING AND TRAPPING

A masochist's day trip
Getting spat at...in the interests of science

I first went ringing at Monk's House Bird Observatory — the 'House on the Shore' — halfway between Bamburgh and Seahouses on the wonderful Northumbrian coast. I'm talking long, long ago. Back in the Fifties. Long before tape-lures. Even before mist nets. Back in the days when we even asked ourselves philosophical questions like: "Do birds mind getting caught?" It was then — and still is — hard to argue that they actually enjoy it, but we salved our consciences — by resolving that it was morally OK for the birds to suffer a little discomfort as long as the ringers suffered more.

I was particularly good at suffering. I caught terrible colds sitting by wader traps in pouring rain; twisted several ankles scrambling up to Starling roosts at Bamburgh Castle; and drowned at least twice, attempting to dive under moulting Eiders at Budle Bay. In all these cases we were absolutely certain the birds came to no harm. We rarely caught any. In the case of seabirds, we were more successful, BUT we always made sure the birds had a chance to fight back. And they did.

Every week a boat-load of ringers would set off to the Farne Islands to ring young gulls, Shags and Puffins. It was a masochist's day trip, if ever there was one. Young gulls are every bit as belligerent as their parents, and twice as ugly. Strangely enough, they even seem twice as BIG, an illusion created by their puffed out, juvenile plumage. It takes courage even to lay a hand on one — or rather two hands, which is what it takes to control the flailing stubby wings, waggling feet and snapping beaks. Ringing young gulls requires skill, strength and considerable co-ordination.

Ringing young Shags requires a clothes peg or blocked sinuses. The stench round a Shag's nest is worse than under a schoolboy's bed. Soggy socks and unwashed underpants are positively fragrant compared to rotting seaweed mixed with droppings and disgorged fish — which, Lord help them, the baby Shags are supposed eat! St Francis of Assisi would be pushed to claim that young Shags are endearing.

They emerge from the eggs looking like misshapen frogs and mature into something rather less lovely than a pterodactyl. They do indeed gorge themselves on regurgitated fish, which they use not only as nourishment but also as a defence mechanism. They invariably threw up over anyone who tried to ring them. Once we'd been soaked in Shag vomit, we felt honour-bound to climb the cliffs to be spattered in Fulmar spit. If we wanted to pay still more penance we went in pursuit of Puffins.

Now, Puffins may well look like portly little old gentlemen in dinner suits but they are elderly delinquents when they are roused. Mind you, I don't blame them. In order to ring them, we did indeed rouse them. The whole business had to be very carefully timed so as not to disturb the birds' breeding cycle. Puffins on the Farnes (as in many other colonies) nest in old rabbit warrens. For a week or so before they lay their eggs, they go house-hunting and in late March they check out the holes before deciding on a cosy one. The Farnes are riddled with these burrows and as you walk across an apparently deserted old warren, you are probably literally treading on several Puffins. You can't see them, but they are down there, considering how they are going to decorate their new home, cuddling their mates, or having a nap.

If you want to ring them, you have to kneel down by the entrance of the burrow and stick your hand down the hole. If you feel nothing, it's almost a relief! If there's a Puffin in there you'll soon know. You'll be lucky to get your hand back. Those pretty little rainbow beaks can snap your fingers off like a pair of garden secateurs. I used to wear a large pair of motorcyclists' thick leather gauntlets. I'd hold the Puffin's body in one hand and the beak in the other.

Puffins are intelligent birds and they quickly realised they were being inconvenienced in the interest of science. They soon calmed down and I was able to release the beak. They'd stare quizzically at me as I shook one glove off to reach for the rings. Then they'd lacerate my bare hand with their feet! Those sweet orange tootsies have claws on them like buckthorns, and they are all the more effective because you expect the attack to come from the other end. After a couple of hours ringing Puffins I looked as if I'd been molested by a werewolf.

But, you know, I bear those scars with pride. They are proof that, in those days, we gave the birds a chance. Don't tell me the lads on Tyneside allow themselves to get duffed up by dark-rumped petrels in return for a drop of DNA!

By the way, much of the above is adapted from my own book Gone Birding; and there's a lovely book about Monk's House — The House on the Shore — by the late Dr E A R Ennion. Both of them have been long out of print. Any publishers out there?

Getting it taped
Is it really cricket in the ringers' nets?

I dare say you are all familiar with the Tyneside dark-rumped petrel saga. OK, we now know they are Swinhoe's. Nevertheless, you may well be wondering what a bird that belongs in Japan is doing spending summer nights in Newcastle. I expect the petrels are wondering that too. All they know is that they were enticed inshore by what they took to be the calls of other night-flying petrels.

But what I'm wondering is, what on earth were a bunch of Geordie lads doing playing petrel tapes in the wee small hours along a stretch of coastline where the average summer seawatch would produce a couple of Herring Gulls and a windsurfer?

Well, of course, there's a simple explanation for that: they were ringers. And ringers will do anything to get a ringing tick. It was worth a try and, be honest, it worked. No doubt, having got lucky with Swinhoe's Petrel, they are out there at this very moment playing penguin and albatross calls. What a pity we don't know what a Dodo sounded like! Extinct? Are we really sure? I reckon it's a fair bet Dodos regularly paddle past Tynemouth on moonless summer nights, if only we could lure them in.

But what about this 'tape-luring' business? Is it really fair? That word 'lure': it even sounds sneaky, doesn't it? Think about it. Just imagine you are a Swinhoe's Petrel, dark-rumped — and dark everywhere else — trying to slip down the Northumbrian coast under cover of darkness, hoping no-one will notice you and ask how come you're about half a hemisphere off course. You hear the cooing calls of friendly mates echoing through the night. "Oh great," you think, "I'm not the only one lost. I'll nip over to that lot and figure out the way home. Or maybe we could discuss starting a new colony. I might even get a bit of late season nooky."

And what happens? You end up being tangled in a net, blinded by a flash gun, and having a hypodermic syringe stuck in your dark rump. What's more, it wasn't another amorous petrel calling to you, it was a bleeding tape recorder. Well, is that fair? All I know is, there's a lot of it about. Tape luring, I mean.

In fact, if you want to witness tape luring on a truly awesome scale, you should visit a certain ringing station somewhere in the south-east of England. I maintain its anonymity only because I know that if I revealed its exact location, ringing tick hunters would flock to it like head-bangers to a Guns 'n Roses concert. Come to think of it, the sound system down there could probably drown out Guns 'n Roses. Only, instead of heavy metal, it blasts out bird song. Electric cables thread between the bushes and reedbeds and every willow tree has a pair of stereo speakers permanently fixed among the branches.

Moreover, the programme has been carefully researched to attract the maximum listening audience. The evening features 'best-loved roosting calls' of Swallows and martins. This is followed by 'night music for night migrants'; while pre-dawn is 'warblers' time' with special tapes for acros, phylloscs and sylvias. (No, I'm not making it up.)

The result of all this aural enticement is that quite phenomenal numbers of birds are persuaded to pop in and have a BTO ring slapped on their legs, rather than carry on to Africa unmarked and unlogged. Now, whether this is entirely in the cause of scientific research or whether it's so that the local ringers can submit their ringing totals to The Guinness Books of Records, is not for me to say. (Though the fact that they are giving away 'Air Miles' to any warbler who introduces a friend to the scheme makes me a little suspicious of their motives.) But again, I ask: is it fair? And what do I mean by that?

Well, heaven knows, I hate to rattle on about how much tougher we had it in the 'old days' but I can't help it, 'cos I'm at that age. And, anyway, it's true. My point is, that birds these days don't seem to stand a chance. The ringer just sits there with his pliers and his tape recorder and the birds get lured to his mist nets, like iron filings to a magnet.

In my day, we had to go out and find them. What's more, we believed that ringing should be just as humiliating and uncomfortable an experience for the ringer as for the bird. My own (one and only) experience of petrel netting wasn't sitting comfortably on Tyneside sea front, with a bottle of brown ale and a parabolic reflector. I was dangling halfway down a precipice in Shetland in a hailstorm and in fear of my life. Swinhoe's Petrel!? We hadn't even heard of 'em. We thought ourselves lucky if we caught a cold!

LOCAL PATCH

My local patch is Hampstead Heath. It is barely a stone's throw from the middle of London and is, not surprisingly perhaps, not all that good. But I'm very fond of it (when I'm not moaning about how dreadful it is). In fact, we — that's me and a couple of other regulars — don't do too badly there. We average over 110 species a year, and get a small share of local — and even the odd National — rarities: Red Backed Shrike, Marsh and Montagu's Harrier, Richard's Pipit — that sort of thing — plus a few I missed, and therefore find too painful to mention (if you want to know more, there's a huge chunk in Birding With Bill Oddie, the book of the BBC series).

Meanwhile...

The joys of birding
In search of peace and quiet

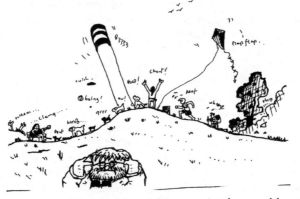

Are my ears getting more sensitive or is the world getting noisier? Or is it just Hampstead, the north London suburb where I live?

Actually, Hampstead — and my bedroom in particular — is usually pretty quiet. But not last Wednesday morning. Long before first light there was a cacophony going on that sounded like someone was playing every record in the BBC sound effects library at the same time. Maybe they were! I didn't immediately leap out of bed and rush outside to look, but I heard the dustmen throwing bins at one another, a builder's lorry dropping a 'skip' from a great height, and a pneumatic driller pulverising the road outside. (Presumably these people do these things in the middle of the night so they can spend the rest of the day sleeping, when it's nice and quiet.)

Added to the human noises, there were cats howling, an insomniac Robin singing, and squirrels carrying out their pre-dawn raid on the bird table. Their attacks are very noisy. They hurl themselves from the trellis onto the hanging baskets — which then swing and crash together, dislodging the geraniums — then take another leap onto the peanut feeder where they break their teeth on the squirrel-proof wire, yelp in agony and fall off onto a pile of plant pots.

By quarter to six all this had me so awake and agitated that even the Balinese wind chimes that I hung up in the garden to relax me began to sound like Big Ben. I decided to get up, get dressed and escape to Hampstead Heath. What do you know about Hampstead Heath? Well, it did acquire some 'notoriety' not long ago as the trawling grounds for — how can I put this? — gentlemen who prefer gentlemen, but that's not why I go there. I love the Heath because it is a fabulous green oasis in the middle of London which attracts a wide variety of bird life. It is a particularly good place for observing visible migration. It might not rival Cape May but it can be pretty lively, especially round about dawn, so off I went.

It was well before sunrise and I expected to be alone. However, I now realise that no matter how early you get out into urban open spaces you will never be there before…dog walkers! There were hundreds of them and, of course, even more dogs. Some owners had as many as six, which probably made up for the ones that didn't have a dog at all, but pretended they had so they had an excuse for wandering round on the Heath at dawn being INCREDIBLY NOISY! I mean, are all dogs hard of hearing or what? Cries of "Here, Rambo!" and "Stay, Tyson!", "Good girl" and "Bad boy", echoed through morning skies at a decibel level that would have drowned out a Madness concert.

The same thing was presumably happening all over the city. From the top of the Heath you can see over most of London. You can hear it too. I'm sure the distant calls were wafting across from Clapham Common and Epping Forest. No wonder the dogs were looking confused. What's poor old

Rambo to do when he's being summoned by six different owners over a 30-mile radius? Hop on a bus and do a tour of all the London parks? (Maybe that's why they call the ticket a Round Rover?).

I stuffed a couple of acorns in my ears and didn't remove them till sunrise, by which time I hoped most of the dog owners had lost their dogs or their voices. No chance. In fact, they were now conversing at amazing volume. (Are these the same people who sit next to me at a restaurant or in front of me at the cinema?).

What's more, the 'soundscape' (euphemism for 'incredible row'!) was being added to by the arrival of the joggers, all wearing personal stereos that 'leak', and heavy breathing like a convention of anonymous phone callers. Talking of which, there were also city gents in pinstripes marching to the Underground, yabbering on their 'mobiles'. Not to mention a couple of Buddhists chanting, a saxophonist practising, and a complete nutter in an anorak walking round and round in circles talking to himself. That was me.

Then, at about 8.30, it all went quiet. The dogs were exhausted, their owners off to find a nice little Hampstead coffee shop where they could shatter the silence, the joggers carted off to hospital with cardiac arrests and everyone else off to work. Including the park-keeping staff. Just when I thought I was alone at last, out came the men in brown uniforms. Private motor vehicles are banned on the Heath. Now I know why. It's so the staff can tazz around all over the place on their tractors, trimmers, mowers, rollers and motorbikes.

Heavens knows, I'd long given up the hope of hearing the peep of a passing pipit (Woody Woodpecker with a megaphone couldn't have made himself heard here), but was a few seconds of silence too much to ask? Well, I tell you, when a rather too handsome gentleman in a white polo-neck sweater (I'm not sure but he might have been a junior minister) approached me and suggested "Maybe we could go somewhere quiet?"...I nearly said "yes"!

Flaming June
The answer is blowing in the wind. Or is it?

12 June, I went birdwatching with an umbrella (to protect me against the elements, not as a companion: I'm not that far gone!) Mind you, it's not an experience I would much recommend. The only occasion on which I attempted to raise my binoculars, to scrutinise what turned out to be Mistle Thrush perched on top of a very distant hawthorn bush and thereby doing a passable impression of shrike, the wind caught my brolly; and whisked me far enough off the ground to convince a gaggle of passing children that I was Mary Poppins. In fact, their spontaneous rendering of 'Spoonful of Sugar' was about the only entertaining thing that happened that morning. Birding on Hampstead Heath in the middle of June is so dull that...well, so dull that you can carry an umbrella and still not miss anything. Well, I don't think I did. To be honest, most of the time the rain was so heavy and my gamp so all-enveloping, that I might as well have been wandering round wearing a cardboard box on my head for all that I could actually see. I heard a Whitethroat, sounding every bit as soggy as I felt, but that was about it. Flaming June! Huh!

There was, it struck me, a certain irony in the situation. As many of you will no doubt attest, most of this spring we had skies so blue and cloudless that all migration passed through or above, totally unhindered and unnoticed by depressed local-patchers below. (One Spot Fly I had on the Heath this May. Just one, and that only stayed for ten seconds, perched on the same Mistle Thrush bush, before launching itself heavenwards in the direction of some more flycatcher friendly habitat, such as Highgate or Golders Green I expect. Somewhere no birder would see it anyway.) As the birdless days dragged by, we Heath-watchers would have pawned our telescopes for a morning of drizzle and mist, with a south-easterly wind. And now I had one. But it was now mid-June; arguably the least thrilling month in the birder's year. Ironical or what? As I trudged homeward under my umbrella, I kept muttering that if only it was late April or early May, or more or less any time in the autumn, we would surely have had a 'fall'. But would we?

The one good thing about June is that it does give you time to browse through your note books, reminisce, anticipate, and indeed, investigate theories. Thus I spent an afternoon browsing through records of the past couple of years on the Heath and attempting to correlate good birds with promising weather conditions. I expected my research to confirm what surely every birder is born knowing: that the east wind, with a bit of the damp stuff, is what forces down the migrants. But, to my surprise, my notes simply didn't bear this out. You have to accept that my definition of 'good birds' is downgraded a bit by the fact that I'm talking about a semi parkland in the middle of London, but I think you'll appreciate my puzzlement when I give you some of the facts. This spring was so poor that I had to regress to 1995 and 1996 to check out a decent day, but eventually I found one. 27 April 1996: good numbers of warblers 'grounded', early Swifts, and a Yellowhammer (well, it's rare in Hampstead). Weather: light westerly, wind, clear and sunny. Mmm. OK then, how about autumn? 10 September: hundreds of hirundines, several Yellow Wags over, a couple of Whinchats and a Redstart. All on a nice grotty easterly breeze? Nope. Due west, and — you guessed it — clear and sunny. 8 October: 5,000 Wood Pigeons on the move, over 100 Linnets — tantamount to an irruption on the Heath, I can tell you. Weather: southerly wind, clear skies. This was getting silly. Surely at least the winter thrushes only appeared on a brisk east wind. Oh yeah? Try 4 Jan: 3000 Redwings and several hundred Fieldfares went over. Wind: south, force two. Weather: pleasant and dry. Q.E.D. Years of self delusion irrevocably exposed!

So where had I got the 'east wind and drizzle' principle from in the first place? Too many birdy holidays on remote islands I guess. Yes, the theory does work on Fair Isle or Lundy, or even on the North Norfolk coast, but I really have to accept once and for all that Hampstead is not governed by the same rules as St Agnes or Cley. In fact, I suspect that at many — or most? — inland sites the wind direction rarely makes much difference. The only rule that does apply, to a point, seems to be that it is worth going out in nasty weather. The few records we've had of waders flying over Parliament Hill are usually when the clouds are low and the air is damp. And 5 September 1995 certainly proved the equation 'orrible weather = good birds, with a large fall of Sylvia warblers and a Montagu's Harrier. On the other hand, the most recent 'real' autumn rarity on the Heath was a Richard's Pipit on 5 October last year, a day with a light north-west wind and skies so totally blue and cloudless that they really belonged to this spring.

Let's face it, the only theory I've come up with is no theory at all. The truth is, especially at an inland site, you never really know. Basically, it's always worth going out. Except, of course, in the middle of June. But then again…what was the date of the Little Bittern that I dipped out on at West Heath in 1995? The 19th of…Flaming June!

The one rare man
Decisions, decisions

Would you want to be on a rarities committee? Any rarities committee: national, county, local. I have been asked a few times, and I have always said 'no'. Well, actually what I've said is: 'not ****ing likely!' The reason I have refused is that I have talked to plenty of people who have been on committees, and I know myself well enough to accept that I couldn't stand the strain, stress, embarrassment and possibly physical danger involved in having to 'judge' other people's records. OK, maybe things are a little less tortuous these days since — thanks to birdline, pagers and mobiles — there aren't all that many rarities whose identity hasn't been verified by a couple of thousand twitchers, several portfolios of slides, and a few cassettes worth of 'video grabs'; but single-observer, notes-only records still exist, and therewith all the problems. I am certainly not going to cite specific cases or names, but the tales are horrendous, involving such dilemmas as: known stringers — 'can't believe a thing he sees'; — attention seekers — 'just wants to see his name in the annual report', wilful deceivers — 'just enjoys causing trouble'; self-deluding fantasisers — 'he honestly believes he sees rare birds'; and even (I kid you not) personal jealousies and vendettas between members of committees and observers — 'he'll never accept one of his records.' Of course, personality is one of the things that isn't meant to enter into the matter. Records are supposed to be judged impersonally, without taking into account who the observer is. Oh come on! Do me a favour. If I get a heard-only fly-over report of a Red-throated Pipit from Mr I. Neverheardovim from Chipping Norton I'm likely to be a wee bit more cautious than if it was sent in by Mr A. McGeehan of Belfast. (And that isn't just because I wouldn't dare contradict Anthony about anything!)

No, rarities committees are not for me. Not ever; never. Or so I thought. Until a few weeks ago I was sort of trapped into breaking my vow. You know I'm always whittering on about my local patch, Hampstead Heath. Well, for a long time it has had its very own annual report (don't ask why, just accept it please). In recent years this has been compiled, edited and published by Mark Hardwick (who also did a similar job for the London National History Society). However, this winter Mark sort of disappeared. (Don't ask why about that either. As far as I know, he's gone to Australia for a year or two.) So what was going to happen to the proud tradition of Heath reports? Would anyone take over the task and responsibility? Who else could do it? Peter, the other Heath regular didn't fancy it. But he did rather like the idea of me doing it. Well, he would wouldn't he? And so it is that — after much deliberation and even more grumbling — I have now become the current editor of the Hampstead Heath Bird Report, and in doing so I have also had to become a sort of one-man rarities committee! Trapped indeed. In fact, it is doubly terrible. Not only do I have no one with whom to share the stress, soul-searching and guilt, the whole thing is made even more insidiously personal by the fact that the only records I have to judge are Peter's and my own! This way I could lose both a friend and my self respect!

So how's it going so far? Well, fortunately, Peter is a very reliable, meticulous and honest observer. (Well, I have to believe that, otherwise I'm going to go completely nuts.) In assessing his 1997 records, I merely suggested that his 'commic' Tern in the first week of April was so early that it was surely a Common, and left it at that. I also felt honour-bound to tell him that the day after he'd had a White Stork soaring high over the Heath there had been a report of a White Pelican circling over Alexandra Palace a few miles to the north. But, then again, there'd also been several other sightings of White Storks about that time, so even if the claim had been sent in by a complete novice it was statistically likely that an enormous black and white bird floating around about 300 feet up would have been a stork rather than a pelican. So that was fine. But it did set me thinking about yet another rarities committee dilemma. How do you adjudicate on 'circumstantial evidence'?

Well, I was about to find out. Consider this. 3 May. A murky morning on Hampstead Heath. Suddenly a large very dark raptor swings into view from over the tree-tops from the north-west. It is being harassed by a single term. 'Female Marsh Harrier!' yells the observer. 'Mobbed by (what was presumably) a Common Tern.' But it's not a great view. Lousy light. The bird is high and half hidden by tree-tops. Not a lot of jizz. Then, to the observer's dismay, instead of circling or gliding in true harrier style, it flaps off directly away and is lost in the mist so rapidly the observer is left wondering if what he's just seen was an hallucination, let alone a Marsh Harrier. He begins to lose faith in his own judgement. Was it just a very very dark badly-seen Buzzard? Oh bugger.

Cut to two weeks later. Same observer meets up with one of the regulars from nearby Brent Reservoir. 'Had anything good at the Brent this spring?' 'Oh yes. Female Marsh Harrier.' 'Oh yes, what date?' '3 May.' 'What time?' 'About 8.30 am. Chased off by an Arctic Tern.' 'Which direction?' 'South-east'. The Brent/Heath connection has been proved many times before: Shelducks, White-fronts, even a Honey Buzzard, all seen first over the reservoir and later over the Heath. Usually about 20 minutes later. The dark raptor had swung over my head just before 9.00 am. Oops, what a giveaway. Yes, I was that observer.

So do I submit the record? Will the committee accept it? But wait a minute, I am the committee. And I've just made my first contentious decision. And I'll tell you something else: I'm having the Arctic Tern as well!

Vive la difference
Common, but fascinating

Wood Pigeons...what's good about 'em? Well, some people find them quite tasty with a crust on top and a spoonful of gravy. Farmers may well enjoy shooting them, but probably not so much for culinary gratification as to stop them nibbling their crops. (The farmers crops I mean, not the pigeons.) I don't think most birders have too many good words to say about Wood Pigeons either. Indeed, I myself have roundly cursed them when I've been lurking stealthily in the bushes in pursuit of a skulky warbler only to be scared out of my wits by a panicking Woodie zipping past my face and clapping its wings like a burst of machine-gun fire. They don't even have the versatility of the good old feral pigeons, who are capable of momentarily raising the adrenaline by impersonating anything from a Peregrine Falcon to a flock of waders. And yet, for many years, birders, especially in and around London, have had good reason to be grateful to Wood Pigeons for livening up many an autumn morning.

Ian Wallace was one of the first to observe the phenomenon and write about it. He saw it over Regents Park several decades ago. I've enjoyed the same experience on top of Hampstead Heath over the past few years. So what goes on? OK, picture this. It is late September or October. Dawn on top of Parliament Hill. There is a great view over London but otherwise it's quite a bleak place. A few people — joggers, dog-walkers, the occasional tramp, and a couple of birdwatchers — not a lot of birds. Until suddenly, out of the gloom to the north, comes a little flock of Wood Pigeons, heading purposefully south. Then another flock and another and another. As the sun rises, the sky fills up. Pigeon to the west of you, pigeons to the east, and more and more from the north. Then suddenly it stops. Within 10 minutes, it's all over but, whilst it lasted, it is truly exhilarating. On the best morning flights, I've counted over 5,000 birds. We also know that similar numbers pass over other high points round London. So where are they coming from, and where are they going to? It would be tempting to assume that it is a massive migration, and this was probably the initial theory. However, and slightly less romantically, it seems more likely that these birds roost in the various woodlands to the north of the city and them disperse southwards into the London parks and green spaces to feed. Instead of returning in an equally spectacular manner in the evenings, they simply filter back slowly throughout the day. Certainly, I have never witnessed a corresponding evening flight. Anyway, the thing is, the invasion of the Wood Pigeons has consistently thrilled me. Until this year. This autumn there aren't any! Well, no more than a hundred anyway. So what has happened to them? Has there been a disastrous breeding season? Is it degradation of farmland? Are Wood Pigeons plummeting like our Skylarks? Or were the huge numbers of continental origin, and has something in the weather patterns this year diverted them away from Britain? Anyone out there got any theories please? I really would be intrigued to hear of similar flights in the past and similar absences this year. Surely Wood Pigeon isn't about to become a red data bird?

Hopefully, it is just one of those temporary mysteries. Certainly Wood Pigeon isn't the only species almost absent over the Heath this year compared to last. Linnets: last autumn we had a record-breaking passage. Flocks of 20 or 30 weren't uncommon. This year I've been lucky to see two or three a day. On the other hand, last year the winter thrushes didn't really arrive in any numbers until after christmas. This year, there were several; good influxes in early October, with hundreds of birds heading west. What's more, this was happening on a south-westerly breeze, rather than the easterlies that have hitherto seemed to be an indispensable condition for Redwing movement. What is going on?

And another thing: have a lot of our Chiffchaffs changed their accent? Heaven knows, I have been birding for 40 odd years and I thought one of the few comfortingly consistent factors was that autumn Chiffchaffs went hooeet. But not any more. It was two years ago when we had a bird on the Heath which uttered a sybillant descending tseeu call, which was nerve-wrackingly reminiscent of a Greenish Warbler, until I got a decent view and realised it looked like a perfectly normal Chiffie. Then last year, when I was in Devon and Cornwall, every Chiffchaff seemed to be calling tseeu. This year it was happening again in Wells Woods in mid-September. I counted 'em: ten tseeus to one hooeet! Have I really been missing this call all my life? Have other birders been hearing it all along? It seems not. Two or three times I've noticed experience bloke's ears prick up, and I've said 'It's only a Chiffchaff', and they've said 'No!' But it was. Again, I sought a little feedback, and, as ever, Ian Wallace was the first to respond. I paraphrase his comments: 'this call seems to be referable to a debatable race/cline known as fulvescens, which probably originates from not too far east'. This may well be, but, if so, they are surely occurring in much greater numbers these last two years. Anyone else noticed them? And why are they here?

Changes, fluctuations, puzzles...sometimes worrying, always intriguing. I guess that's what keeps us birding.

Is it me who's falling apart?

When 'wet' doesn't actually mean damp

I have just returned from my daily amble round my local patch: Hampstead Heath. It was, as it happens, a deeply birdless experience. Never mind, it was still a glorious winter's morning, with a sparkling sun burning off the frost and the early mist. It was also totally windless; all in all, the sort of day that you can be absolutely sure there really is nothing to watch. But, like I said, never mind, I am not so totally obsessed with birds that I can't appreciate a nice bit of weather.

I enjoyed my walk. Except for one thing. By the time I got home, my feet seemed to be wet. Now I say 'seemed' to be wet, because they couldn't really be, because my flashy new boots are lined with that waterproof stuff that ends in 'ex' and is guaranteed to keep your tootsies dry even after a couple of hours tripping blithely through Hampstead's dewy meadows. And sure enough, when I kicked my boots off, my socks were not really damp. They just felt it. My feet were sort of cool and clammy, yes, but not actually wet. But they might as well have been. 'Cos they didn't feel dry.

So what on earth am I getting at this month? Well, own up, we birders are pretty hung up on 'gear' aren't we? It might be waterproofs or it might be optics. I just want to know…am I alone in finding some of the maker's' claims a little less than totally convincing? Let's take this waterproof business. In fact, I can't really grumble about my new boots. The inside may have this 'ex' stuff lining, but the outside is all soft and suedey, like a track shoe, and is, I imagine, about as water-resistant as a sponge. My feet are wonderfully unblistered but, if I'm not prepared to wear crippling leather like those butch chaps who splash around in the adverts, I surely can't expect them to remain completely dry as well.

But then, of course, they were dry. They just felt wet! In the same way that my knees always feel wet if I go out in the rain in my very expensive, supposedly waterproof, lightweight, non-rubberised, breathable trousers, made by a widely-advertised company who specialise in trendy outdoor clothing. I even took these back to the shop. I was told I must be hallucinating. "Our trousers do not leak. We shall pour a kettleful of water on them to prove it." And they did. "If your knees feel damp, it must be coming from the inside, not the outside." So I must have leaky knees. Anyone else have that problem? Or is it just me?

Likewise, how about my £200-plus anorak, made of another miracle material (probably also ending in 'ex')? It's not waxed, so I don't suffocate, drown in my own sweat, or get mistaken for a 'royal'. It does take a week to dry the outside, but it never gets wet inside — except in the pockets. I asked the man in the shop about that too, I should have known what he'd say. "Ah, well, yes; it may feel damp inside the pockets." But it isn't really? "Exactly." Just like inside my socks. And apparently the pages of my notebook are only pretending to be soggy, and the ink only looks as if it's running, and my lens cloth is only doing an impersonation of a wet-wipe!

Which brings me to spectacles and eye-cups. I've been wearing glasses for birding for about a year now. I rapidly passed through my contact lens phase. Heaven knows, I'm not blaming the miraculous little lenses. I admit it's entirely my fault that my eyes are so small and my sight so inadequate that it takes me half an hour to get lenses in, and half a day to get them out, leaving me a bloodshot nervous wreck. In fact, I almost don't mind wearing glasses (I definitely see more than when I'm not wearing them!) but, let's be honest, using binoculars is easier with the naked eye. Which is why, for the first nine months, I pushed my glasses back onto my forehead. This worked pretty well, except when birds flew directly overhead, when my glasses fell off, or deep in woods, when I kept losing vital seconds and several birds. So then I decided to try the 'keep 'em on' approach...

Well, I don't care whether the eye-cups fold back or click in, or how many millimetres the 'eye relief' is (and what the dickens is 'eye relief' anyway?). In my experience, looking through 10x binoculars and spectacles you feel like a 'peeping Tom', peering through a couple of knot-holes. I have to admit, friends who've worn glasses all their life say they don't feel like this, but I say what you've never had you don't miss! "Retains full field of view" — I don't think so. Or is it just me? I suspect it is. Probably my eyes are as faulty as my knees and feet. Fortunately, I have arrived at the perfect solution. For maximum magnification I use 10xs and push my glasses back. For maximum field of view, I switch to 7xs and keep them on. And for the ideal compromise, it's 8xs and a bit of both. Expensive business, this birding, isn't it?

READERS LETTERS

In case you're wondering about those mysterious Chiffchaffs...I actually received a lot of letters from birders, from various parts of the country, saying that they too had noticed this 'new' *pseeu* call, and agreeing that it seemed to be getting commoner. I also got a very authorative letter stating that it was simply a recognised call of juvenile Chiffchaff, and implying that it had therefore been uttered since time immemorial (or since the evolution of Chiffchaffs anyway) and that I — and the letter writers — simply hadn't noticed it before. I found this hard to believe; but then I also found it hard to believe that we were witnessing the appearance of a new 'race' or dialect. But — to make things even more puzzling — the *pseeu* sightings (or rather hearings) peaked throughout the years 1995 and 1996, but then diminished in 1997 and 1998 In fact, to really bring home the point, I witnessed a huge fall of Chiffchaffs in October 1998 (see 'Do the Right Thing' page 43) but I heard not one of them give a *pseeu* call. (And yes, plenty of them were undoubtedly juveniles.) So...I'm still baffled. But, like I said, that's one of the things that makes birding so endlessly fascinating.

If nothing else, the Chiffchaff episode confirmed that there were readers out there capable of putting pen to paper, but back in 1994 I was rather less than inundated with correspondence, as the next piece indicates...

Is anybody there?
Please write and let me know

This is my 'first anniversary' article. I've been writing for *Birdwatch* for a year. How time flies when you've got deadlines to meet. In several of the pieces I've done I've sort of invited readers' comments or, at least, I expected to get letters. Well, either the Editor has been protecting me from poison pens, or else *Birdwatch* readers have got something better to do than write...like birdwatching perhaps? I have had a few people write to me, so I thought this month I'd just do a sort of update on readers' responses.

The first one is actually an interesting negative. I wrote a piece about birds that got away, telling the story of what I think was probably a Pintail Snipe in Scilly some years back. I expected all sorts of claims and confessions to ensue. But no. Maybe the memories are just too painful to relate. At least I like to feel I might have inspired whoever it was kept seeing a Pintail Snipe in Ireland early this year. I suspect that one's going to 'get away' too.

Tape luring and ringing. A few letters about this one. It seems there's quite an anti-ringing lobby building up out there. I note particular sarcasm directed at the increasing phenomenon of queues of twitchers waiting by mist nets for the latest trapped rarity to be paraded 'in the hand'. I've also had several letters from Blyth's Reed Warblers who found it totally humiliating. They have vowed to hybridise with any 'acro' who takes their fancy and form a superspecies, so all listers will lose at least half a dozen ticks. You have been warned.

Anoraks and spectacles. This is actually the one that got most feedback. I had lots of sob stories about leaky materials with an 'ex' at the end, and one with an 'ic'. As is happens, I have since bought a new coat — with an 'ex' — that really is waterproof. It does, however, have luminous purple shoulder patches, so that I have to avoid Oxford Street, otherwise I keep getting mistaken for a French student and Hare Krishna people try to sell me love and paperbacks. (If you have never been in the West End of London and have no experience of this sort of thing, think yourself lucky.)

As regards using bins while wearing glasses, this produced a genuinely revelationary postcard which, alas, I've now lost, so I can't give the bloke due credit for changing my life. But I can now pass on his advice. Spectacle wearers note this, and note it well: it's not the bins, it's the specs that make the difference. Big spectacles push the bins further from your eyes, and this is what gives you the 'through a couple of knot holes' type of effect. Get a pair with small lenses, that fit as close to your eyes as possible and you really will get a full width of view, when you turn back or push in the eye cups.

I tried it with a pair of old John Lennon-type NHS 'granny glasses' I had in a bottom drawer and it really does work. They also look ever-so hip, 'cos they fit in with the current back-to-the-Sixties fashion. A pair of loon pants and a CND badge complete the effect.

And that's about it really. I'm still awaiting the backlash from dog owners after last month's article. I've been sent several Rottweiler-shaped packages but they go straight down the waste disposal unit.

Meanwhile, I'm going to conclude with two perfectly serious pleas for 'feedback'. First, can someone please explain to me what 'bird days' are? You must have seen the expression in annual reports: "Pied Flycatcher: 43 bird days in September". I used to think it meant days on which the species was recorded, but it can't be that. After all, '30 days hath September,' not 43. so is it the number of birds added up — in which case, what does the 'days' bit mean? Or is it birds multiplied by the number of days? Or what? Please explain it to me, or stop using the expression.

And finally, an authentic ornithological puzzler. At the end of last year, I saw a gull on Hampstead Heath playing fields that simply 'did not compute'. Basically, it looked like a sub-adult/second-winter Common Gull. Except that it was rather petite — not much bigger than Black-headed Gull — with a more delicate head shape, a slightly paler mantle than typical Commons, and — get this bit — a dark reddish bill! My 'logical' — yet unlikely? — conclusion was that it was a Common x Med Gull hybrid. I'd never heard of such a thing until — by amazing coincidence — *Dutch Birding* dropped through my letter box the next day. It included a 'mystery bird' that had been identified as a 'possible Common x Med Gull' (first-winter). If you take *Dutch Birding* (and you should), have a look at the photo and imagine that bird a year later, and that's what I saw on the Heath. So please — pretty please — has anyone else seen anything like that? Don't be shy, we won't accuse you of hallucinating or drinking too much — you can even remain anonymous if you like — but do get in touch.

Correspondence Closed
At least one mystery cleared up

OK folks, it's postbag time. A couple of months back I appealed to readers for enlightenment and elucidation. I received quite a few letters, though most of them have been buried under the deluge I got after doing the BBC's Bird in the Nest series. Boy, was that popular! Avian soap operas, is that what the viewers want? Great. Move over Neighbours, I'd say. Now, how about an Australian Bird in the Nest? That should really top the ratings. "Yes, those kinky kookaburras are at it again!" Seriously though, I must say I did enjoy doing that programme, but I have to face the fact that I shall be getting polaroid pictures of Robins in toilets for the rest of my life. Never mind. Very satisfying. Meanwhile, back to your letters.

Some of them were merely appreciative, including one from somewhere foreign — where is Troisdorf? — so thanks indeed for those. I shall carry on writing, and the Editor of *Birdwatch* is thrilled to know the mag is still migrating successfully. I only had one response to my mystery gull appeal (remember the possible Common x Med hybrid I saw on Hampstead Heath, accompanied by a photo of a similar bird from *Dutch Birding*?). This came from a couple from Cumbria who saw something very similar up there. So the conclusion is...these weirdo crossbreeds probably do exist. So, hey twitchers, why not seek them out and tick 'em? They're less likely to have come out of a cage than half the stuff on your lists these days!

I'd also appealed for an explanation of the term 'bird days'. "What does it mean?" I'd asked. And answer came there...one! Is that because the truth is that all you local report editors have been using this expression for years without really understanding it? "Well everyone else uses it, so I suppose we better had do, too."

Anyway, thanks to John Hopkins of Exeter for solving the mystery. I quote: "The number of bird days in a period is the sum of the daily totals for that species during the period." (Following this?) "So, for example, if during a three-day period the number of Firecrests recorded in your garden were two, nine and three respectively, this would give 14 bird days." (Got that?)

John then goes on to say that the system is "often misused and misleading". I'd agree. If I claimed "14 bird days" and what I really meant was "recorded on three consecutive days, with a maximum of nine," I'd be bemused myself. Especially as I've never seen a Firecrest in my garden anyway. So to anyone thinking of using the bird days method, my plea is: think again, 'cos no-one understand what it means and those that do think it's pretty useless. No doubt several scientists are reaching for their pens at this very moment to disagree, but please don't. Further correspondence on this matter is — as they used to say in BB — closed. Well, it is to me, anyway.

One more bit from a letter, though. This one comes from Michael Blencowe of Plymouth. Mike confesses to having lapsed as a birder a few year ago, basically because he couldn't take it seriously any more, precisely because other birders were getting too damned serious. He began to have doubts when he attended the scene of the Plymouth Sheathbill (remember that one?). He writes: "I have since viewed sheathbills in their native habitat (in Patagonia)...those birds don't fly more than three feet if you kick them (not that I spent my time in South America kicking seabirds, of course). So why people were getting excited about its appearance here I don't know. I was more intrigued at how it had made the journey from the Falklands to England aboard a warship without ending up in the company of roast potatoes and three veg!"

Mike goes on to remember "the Olive-backed Pipit which was literally in someone's small terraced back garden. I can still recall the expression on that woman's face when she opened the curtains in her underwear to find 50 telescopes pointing her way. I sometimes blame this experience — of seeing a scantily-clad young woman — for showing me there was more to life than small brown birds." Be grateful, mate, I'd say. I've never seen a woman with curtains in her underwear!

Michael does add that he has recently returned to birding, but clearly with a lighter touch, a saner attitude and a sense of humour. Good on him. I thank him for his letter, both for giving me a chuckle and for making me review my column (Oops missus! Sounds a bit naughty). What I mean is...I've been writing in *Birdwatch* for over a year and I'm beginning to run out of topics. For a start, Anthony McGeehan keeps covering things quicker and more eloquently than me (it's these Irish you know: talented and poetic with it), and now my own readers start writing funny letters.

Obviously, I need a change of direction. So, from next month on, I'm going to be delving back into my notebooks and attempting to recapture some memorable birding moments. I shall call them 'Gripping Yarns'. Tee hee. I only hope they are. But if they're not, please don't write and tell me!

Dear Bill

From fan mail to fantails

I get letters. When my Birding series was on the telly earlier this year I got lots of letters. Literally, hundreds of them. I replied to them all. I've just finished! A bit of a chore I admit, but enjoyable too, especially since every single one of the letters was basically 'nice'. In fact, I was almost disappointed not to receive anything from my long time anonymous poison-pen pal who has been periodically insulting me for nigh on 30 years. I guess he must have snuffed it, or transferred his vitriol to Noel Edmonds. All he'd have to do would be to recycle his jibes: 'Your long hair and beard make you look like a dirty scruff. You fungus face. Can't you afford a shave? You need a dose of the army. You millionaires are a bloody disgrace.' Yes, well, I'm sorry, but I haven't time to shave 'cos I've got hundreds of letters to reply to; and I need to be quite well off since barely one in ten enclose a stamped addressed envelope (future correspondents please note). Never mind, it's stationary well bought and stamps well licked as far as I'm concerned.

The letters fall into about half-a-dozen categories. Probably few of them are from what one might call 'proper birders', i.e. readers of this magazine, which is one reason I can't use this column to publish half-a-dozen 'standard replies'. The other reason is that I feel that most of them truly deserve a personal response. Especially the ones in the first category. 'First' with me because it is undoubtedly most gratifying. These are letters from people who are largely housebound, either through old age or illness, but enjoy vicarious birdwatching via the TV. They write to thank me. Believe me, I thank them. That's real job satisfaction, I can tell you. Category two is equally pleasing: people who write to say they had never tried birding but have taken up after seeing the programmes and have discovered a whole new element of life. 'Brilliant!' as the bloke on the Fast Show would say.

The rest of the letters are generally less emotionally affecting, but arguably even more entertaining. Although there are just a few who do tempt me to facetiousness. Like 'Dear Mr Oddie, where can I get a book about ducks?' Er...a book shop? Plus, there are those that look as if they have been written by an inebriated spider with a magnifying glass (at least it could have used a pen! Bum bum!). Fifty ways to be illegible. Nevertheless, I try to decipher them and usually discover that, like the rest, they display a concern and involvement with birds that, though it may sometimes be sentimental or even naive, is still utterly genuine. We shouldn't knock it. OK, we heavies may be absorbed with tertials and taxa, but I presume we are all basically into birds because we like them!

So what are the other categories concerned with? Blue Tits. Lots about them. Either joy at record fledglings: 'Ten babies left the box. Is this a record?' Like, I should know! Or sorrow at massed bereavement: 'We looked in the box and there were six dead chicks. What did it?' I don't know that either, but my motto is 'when in doubt, blame the cat' Talking of which, I am accumulating quite a collection of snapshots of cats watching Birding on the telly. Well, I suppose it's the feline equivalent of Master Chef. Which reminds me, why don't pet food companies advertise cat-food as having 'Blue Tit flavour'? Of course, they'd have to draw the line at contains real Blue Tits', but, let's face it, cats don't really go for that 'liver and marrow bone jelly' nonsense; it's that 'unmistakable taste of tit' that would really get Tiddles slavering. (Not just tiddles. Ed.)

And talking of bird losses, but seriously, that's another disturbingly large category of letters: 'Where have all the ****s gone?' Fill in the blank with any number of species: Song Thrushes, Blackbirds, Sparrows, House Martins, Swallows, Lapwings etc etc. Believe me, the folks out there don't need the BTO to tell them that birds are in big trouble. They notice, they miss them and they care. And they ask me: 'What can be done about it?' And that's another question I can't answer.

Which brings me to the final, and favourite, category: the 'mystery bird' letters. They invariably begin: 'Dear Bill Oddie, I have seen this funny bird'. Then they describe it. Very occasionally there is a drawing that often barely looks like a bird at all, let alone a particular species. The letter always ends with 'I can't find it in the book(s). Please can you tell me what it is?' Now, I could write back with various glib responses, such as 'No!' Or, 'It's a Jay.' Or, 'It must be an escape.' Or, 'Obviously it's a first for Britain, put it on Birdline immediately.' But I am much more conscientious than that. I really do try to picture the bird from the description, or decipher the drawing and come up with something likely. But I don't know why I bother. The honest truth is that of the 20 or so 'mystery birds' that came in the last batch of letters I was able to identify not a single one. I don't think it's my incompetence. Neither it is really 'incompetence' on behalf of the writers. It's just that, well...they are not proper birders! Well you have a go. What's this then? (I quote) 'Size: as big as a large male Blackbird. Colour: pure white with a black cap, but when it flew off the body but not the breast and the underside of the wings were bright orange/red.' Mmm. So it changed colour after take off? And how about size variation in male (but not female?) Blackbirds? That surely merits a note in BB. I evoked the 'exotic escape' clause on that one, but maybe you recognise it immediately. If so, do write me a letter.

Or maybe not.

Some you win
What really turns me on

Blimey, I dunno. Last month I asked for a little feedback and what do I get? Poison-pen letters! First of all, Clayton Jones of Hampshire accuses me of being a transvestite: 'Bill Oddie reminds me of my mother-in-law.' What? You mean she's short dumpy and bearded as well? Then he really insults me by questioning my commitment. 'Come on Bill', he asks, 'Where's the old enthusiasm gone?'

Well, at least it made me ask myself: 'Has the old enthusiasm gone?' The answer is, 'certainly not'. It might well have got a bit older — and wiser? — but it's definitely still there. But where is it? Well, I'll tell you one things, it's not stuck in the past, which is what I rather suspect I was being accused of. This is partly because the past really doesn't exist any more. Some of the situations that really turned me on when I was a 'twenty-something' birdwatcher (which I assume Clayton still is) have more or less gone. Twitches that numbered in tens rather than thousands, the chance of finding your own rarities, the possibility that you might not recognise them because the books weren't good enough. All that naive, pioneering, exciting stuff that Ian Wallace writes so brilliantly about. The days of innocence. I'm glad I was part of them, but they are truly no more. And maybe that's why I sound a little mother-in-law-ish sometimes. We old codgers love a bit of nostalgia, but you mustn't revel in it.

But perhaps there's another reason I'm not permanently leaping about with manic excitement these days. In recent times, I've put every bit as much of my enthusiasm into conservation as I ever did into trashing the twitchers in the Little Black Bird Book. Energy rather better spent, I'd suggest, actually. The trouble is, though, that it's not always a jolly experience. Habitat loss, pollution, poisoning, and politics aren't exactly frivolous topics. 'We all know it's not a perfect world out there without being reminded of it' writes my pen pal Clayton. Well, if you're involved in conservation, you are reminded of it every flipping day. What's more, it's part of the job to remind everyone else about it too. I know I'm using a cheerily meant ribbing to bring up a serious point, but that's how I feel today. The thing is, sometimes conservation seems like a series of losing battles. And it can get you down. But when you win one, it feels terrific.

Which brings me to Cyprus. For many years the islands had an unsavoury reputation for the slaughter of migratory birds. Liming, netting and shooting. I'd heard about it and felt incensed, but that was nothing to the emotional outrage I experienced when I actually witnessed it. Standing and watching bee-eaters, falcons, wheatears, warblers — anything and everything — writhing in their death throes, or blasted from the skies, leaves you feeling sick, and angry, and appallingly powerless. But people aren't powerless. Over the past 10 years, I've done all I could to help 'Stop the Massacre'. Many people have done much, much more. Slowly, things improved. First, the illegally imported mist nets were largely confiscated, then the lime sticks became scarcer and scarcer. Then, in 1991 the Government announced a ban on spring shooting. Last year, April 1993, I co-led a party of birdwatchers to the island, only to be appalled when the new Government re-introduced the shooting. Once more, the campaigns leapt into action: Friends of the Earth, BirdLife International, RSPB, petitions at the Bird Fair, and so on and so forth.

The response was tremendous, the pressure on Cyprus must have been terrific. And yet I had one overriding worry. There were activists in Cyprus itself, but undoubtedly most of the protests were coming from 'outside'. It seemed like much of Europe was in a rage, but what of the Cypriots? Rumour had it that a high-up minister on the island had stated that they would not be pressurised by 'a bunch of foreign nutty extremists'. I hated to hear that, but I sort of understood. Then came a master stroke. A long-time campaigner, resident on the island, instigated the idea of a poll amongst Cypriots to gauge their attitude to the spring shoot. BBC Wildlife funded the operation and, late in March, the results were published. Over 80% of the people were strongly against the spring shooting and, amazingly, over 70% of hunters agreed that it was undesirable. It was the perfect response to that minister. Was he calling his own people 'nutty extremists?' Early in April, I receive a letter from Cyprus: 'in the light of the wishes of the people, and outside opinion, and in accordance with the Berne directive on migratory birds, the Government had announced a ban on the spring shooting'. We still have to discover if the ban will be permanent, and if it will lead to further conservation measures but, believe me, hearing that news was every bit as thrilling as nailing a Pallas's Reed Bunting.

So, do you still wonder where my enthusiasm has gone? Well, try Cyprus in April for a start. On Paphos headland, early morning. Flocks of Red-throated Pipits and flava Wags. Short-toed and Bimaculated Larks. Ortolan and Cretzchmar's Buntings. Rüppell's and Orphean Warblers. And five species of wheatears, including the island's first Mourning, which our party found, and is now attracting the island's first major twitch. That was 1993. I bet 1994 was even better. That's where my enthusiasm is. Pity I'm stuck in London writing this article!

BIRDING ABROAD

And talking of Cyprus...this section begins with a little controversial bird incident which sort of carries on the correspondence theme, in so far as it took a couple of consecutive pieces to resolve it.

The rest of the section covers all sorts of foreign affairs...

Tour of duty
Overseas birding has its drawbacks

"There is a corner of a foreign field that is forever England"...especially if it's got birders in it. As I write, it is mid-April and I have just returned from leading a group for a week in Cyprus.

Of course, when it comes to anglicisation Cyprus has a considerable head start. For a kick off, the British army has had a presence on the island for a long time. It was once — and possibly sometimes still is — a blessing, but these days it is also all too often cause for considerable embarrassment. So too are many of the signs of Brit 'culture'. Within chundering distance of our hotel in Paphos were various pubs and nightspots meant to remind us of home — the Queen Vic, the Rovers Return, the Drunken Duck (specially for the birders?) which offered a telly with "live footboll" (even spelt with a Geordie accent!), plus others that had got it a bit wrong: the Chippendales Bar (surely those glistening jessies aren't British?) and the Lucille Ball Cafe (eh?) and, saddest of all, Mr T's Top Burgers, boasting "guaranteed British Beef" (hasn't anyone told him?). Yes, surveying the Cyprus scene certainly makes you proud of the old Union Jack...not.

But what about the birding situation? Well, frankly, it's getting a bit crowded. Early mornings on Paphos headland in April are like a strange amalgam of Portland Bill, Scilly and the Rutland Bird Fair. There are likely to be several 'official' tour groups, some of them only half a dozen strong, some of them pretty enormous. We had 22 in ours! (Pause while I scream at the memory of trying to guide 22 people of very mixed ability on to a skulking Cretzschmar's Bunting. Apparently I failed, since several clients admitted some days later that they "still needed Cretz.")

Then there are the 'freelancers', doing their own thing — the particular thing often being a clue to their region of origin. For example, show lads from Hartlepool a headland and they'll seawatch from it, even if it is still half an hour before dawn and there's nothing to see but a couple of silhouetted Yellow-legged Gulls half a mile away (quite enough to provoke a 10-minute discussion on 'races' and 'wing-tips'). Once the sun is up, it is likely to illuminate several familiar faces from the UK scene: photographers photographing, artists arting, and twitchers twitching. They look and sound like experts, because they are. Others look like experts but definitely are not. It seems that on the headland the typical British tourist has forsaken his 'Kiss Me Quick' hat and stick of candy floss for a pair of Leicas and a Kowa on a tripod. Yes, at Paphos birding is for everyone. There are even entrepreneurs sidling up and furtively slipping you the number of the local Birdline as if it was as titillating as a 'dirty postcard'.

But surely there is nothing regrettable about this birding escalation, is there? Well, I don't think so...but I did talk to one expatriot local birder who seemed a little resentful. "It used to be wonderful here, " he said wistfully. "Masses of migrants, and just a few birders to share your enjoyment with. If you found a rarity, you probably had to be your own judge and jury. Now everyone's under scrutiny. It has become incredibly stressful." I wasn't entirely sure what he meant until I saw that wheatear.

Early one morning I was called over by another birder. "Male Finsch's Wheatear on those rocks." And sure enough that was what it seemed to be. A truly stunning bird. The white bits dazzling like snow, the black parts like velvet, and — crucially — no pale gap between the face and wings. That's what distinguishes Finsch's from the almost equally striking and very similar Black-eared Wheatear. A few other people saw the bird. It made us all happy, then it disappeared.

If only it had stayed that way. Alas, I had the good fortune to relocate it that evening, hiding in a quiet stony field 'round the back'. Heaven knows, it has probably been trying to escape controversy, but it failed. More people came to inspect it. By the next morning there was muttering in the ranks, and a move to demote the bird to a "funny Black-eared". Except that it wasn't making anyone laugh. There were those who had noted a very thin gap between the black face and wings when the bird stretched. They also pronounced "too much white on the tail", though this verdict depended on which field guide you consulted. There were others who took the first impression jizz approach: "Never mind the feather waffle, this bird is saying 'I'm a Finsch's'." What the bird was probably saying was "Pardon me for living" and "Get me out of here". So was I. I heard voices raised, oaths uttered, blows almost threatened, and even Lars Jonsson's accuracy questioned. Stressful? I should say so. I knew what the expatriot meant. Just like home.

So what are you thinking? "Hmm, maybe I won't go to Cyprus next spring after all"? No, don't say that. The birds are still brilliant, and it is vital that 'green tourism' thrives there and encourages conservation. Or maybe you are thinking: "I'll try a different part of the island, or a different time of the year." Well, that's not a bad idea. But of course what you really want to know is: was it a Finsch's or not? Search me. For what it's worth, the little display posture drawings in BWP show a distinct gap when the bird is contorted; but it did have a little more white on the upper tail that Lars shows (but has he seen a male in spring?). I can't find a picture of Black-eared that looks like this bird did. But then I can't find a photo of a male Finsch's either.

One thing I do know, though: it was a very beautiful bird, and I can't help pondering that that is what made it worth seeing, not its name. Tick of not, it didn't deserve to be 'rubbished'. But that's a very un-British attitude, isn't it? Sometimes, I don't know where I belong.

Got it sussed...not

The great wheatear controversy continues

You remember that wheatear I was telling you about last month which I saw in Cyprus in April? The one that could have been a male Finsch's Wheatear, but on the other hand might have just been a 'funny' Black Eared?

It is possible that you don't care a chat's fart either way, but I do. What's more, I feel I have to head off accusations that I've gone soft in my old age. OK, so what if I have drooled over Blue Tits on Bird in the Nest, I can still bandy tertials and primary projections with the best (or worst) of them. Own up, there is nothing I like better than a good old abstruse and tricksy ID wrangle.

And they don't come any tricksier than wheatears. Cyprus in spring is full of them. It's quite interesting to ponder the practical approach in the field. Do you rely mainly on plumage or jizz? Take Isabelline for example. Until you see one, approximately every 10th female Northern strikes you as suspiciously pale, or large, or upright. But that doesn't make it an Isabelline. Remember those old birding adages: "If you are not sure, it probably isn't," and "When you see one, you'll know." And indeed you do know Isabelline when you see it. But why? OK, the stance is a bit different, but really it's that beige cloak that is the give-away, extending right down over most of the wings (there are no discernible coverts really). You just don't get Northerns like that. Once seen, never forgotten.

Then there's the Black-eareds. In Cyprus, they come in a quite bewildering variety. Check out Lars Jonsson's plateful. All them, and more. Of course, it's the females that are 'hard'. You could go into pages of feather waffle, but there's hardly two that look alike. So you tend to go for the jizz: smaller, chat-like, given to perching like Whinchats too. Plumage-wise, most are obviously not Northern, but some almost could be. Even the supposed greater amount of white on the tail doesn't always work — it's very variable on females. But surely the males are no problem?

Aha...now we're approaching the nub. Black-eared Wheatears belong to two races: hispanica (west) and melanoleuca (east). Both occur in Cyprus. Western birds are basically sandy, while eastern ones can be almost white, except of course for the black bits. The black 'ears' can be just that — ear coverts — or they can extend to a black throat as well. Already, therefore, there are several combinations available, especially as the tone of the pale areas is highly variable. However, they do all look like Black-eared Wheatears.

Except the one that looked like a Finsch's. It looked like Finsch's because it was rather robust, basically almost pure white, and the black of the face appeared to be widely joined to the black of the wings. Two points, however, were a little worrying. There seemed to be a lot of white on the tail, biting into the terminal band — a Black-eared feature — and, just now and again, if the bird twisted its neck round, a very slight gap appeared at the shoulder. Thus, as I recounted last month, the investigation began. What was fascinating was how it progressed.

First, there was the book consultation. Most field guides were available, but the ones that seemed to carry most authority were the excellent Macmillan *Birder's Guide to Europe* and the Middle East, and of course Lars Jonsson's *Birds of Europe*. In both books, the illustration of Finsch's looked closest to our bird. Moreover, though Macmillan showed quite a wide black tail band, Lars's was somewhat narrower. Neither book carried anything but a cursory warning of "possible confusion with Black-eared". The money was surely still on Finsch's.

Then we spoke to an ex-pat birder who had experience of both species. "How do you identify Finsch's?" we asked. "Easy. Finsch's is the only wheatear we get in winter in Cyprus, so that's how we recognise them!" He chuckled. We winced. "But surely they can occur on passage?" "Maybe." "So what's distinctive about them?" Well the black of the scapulars almost joins over the mantle, until it flies, when it shows a narrow white flash right down the back."

Just like the one at Paphos. Our money was still on Finsch's. But our friend seemed sceptical. Rightly so.

Salutory conclusion coming up. The bird was also seen by an eminent British bird artist. He admitted that he had not seen Finsch's before, but was happy — nay delighted — to be told that this must be one. On a blustery evening he made sketches, and the next chilly morning too. In these drawings the bird looks stocky and the black joins. Then a strange thing happened. As the day warmed up, the bird got slimmer, and slowly revealed a gap...and its true identity. Apparently, being hunched up against the cold had affected both its posture and its plumage. It was a Black-eared.

I was told this news back in Britain. But in case you're thinking it's clear cut and easy, I have also consulted BWP and a video of Cyprus birds, both of which clearly show that Finsch's can show a narrow gap in certain postures. Nevertheless, it was a Black-eared.

I also spoke with another Middle East regular who described the jizz of Finsch's as "more upright and short tailed, a bit Isabelline-like". As with that species, I suspect it's a matter of "when you see one, you'll know". OK it was a Black-eared.

PS I have just been told by the Editor of this magazine that Black-eared and Pied Wheatears have been known to interbreed. So what do those offspring look like? And what's to stop Finsch's from joining in the fun? Or are hybrids the last last resort of the incompetent?

Write out 1,000 times: "It was a Black-eared."

Accentor-ate the positive
Barcelona, spring 1962

OK. So it's 14 November and my article is due. I'm thinking: "Shall I write about Alpine Accentors?" While I'm making up my mind — and to avoid actually getting down to work — I ring Birdline. First bird up: "At Rimac, in Lincolnshire, an Alpine Accentor". Weird or what? It is surely a sign. So I will write about Alpine Accentors.

Not that I'm exactly an expert on the subject. I've only ever seen one. It wasn't in Britain and it wasn't in the Alps either, and it was a very long time ago. I have, mind you, tried for them rather more recently. This May, in fact, while leading a party in the Pyrenees. We didn't see any. Probably in the wrong mountains. After all, they aren't called Pyrenean Accentors are they? But they are supposed to be in the Pyrenees, and we didn't half try hard.

We had lots of 'sites'. Most of them were ghastly ski resorts that resembled a hybrid between Blackpool pleasure beach and a rubbish tip, but which were set against snow-capped mountains that looked so unreal they seemed like a back projection. You expected Julie Andrews to come frolicking through any minute. You also expected Alpine Accentors. But there weren't any.

The final site was probably the most hurtful. It was somewhere on the Spanish/French border, in a snow-field spangled with orchids and miniature daffodils, with Chamois scampering over the crags and Rock Thrushes dancing around the boulders. My co-leader even assured me that last time he'd been there (a few years before) there had been several accentors flitting around on the very scree slope which was at that moment crunching beneath our feet. But there weren't any that day.

"So where the **** are they?" I enquired. "Er...maybe higher up?" suggested my co-leader. "The last time (don't you just hate that phrase?) one of our party climbed right up the top of that ridge and yelled down: 'It's brilliant up here, there's Alpine Accentors singing everywhere'." So, muttering that history could repeat itself, off I went, scrambling upwards like a demented Marmot. The only satisfaction was that I reached the top and no-one else even tried. Least of all my co-leader. Probably because he knew there was naff all up there. I frightened a Fox, startled a Chamois and started a small avalanche, but I enjoyed not a sniff of an accentor. Well no, that's not quite true actually.

There were in fact three 'accentors' singing up there on the ridge that day. Would you believe Dunnocks? (I'm blowed if I'm calling them by their more pretentious name if they're going to mock me like that.)

So, be warned, it's perfectly possible to spend a week in the Pyrenees in May and not see an Alpine Accentor! I admit I was sort of narked. But then, how thrilled would I have been if we had ticked them off? I do sometimes find this 'sites' business a bit uninspiring. I mean, you can be pretty sure birds are there...somewhere. If you them, it's satisfying, but still sort of predictable. If you don't, it's downright depressing. It's like going to twitch a long-staying rarity and arriving on the day it decides to push off. How much more exciting when you have no expectations.

Which brings me to the Alpine Accentor I did see. Talk about bizarre circumstances. It was 1962 and I was on a college rugby tour of Spain. No, I didn't know they played rugby in Spain either, and back then they probably didn't. So, why were we touring there? I dunno. Maybe we hadn't had a very good season and we wanted somebody to beat. Or maybe we were on a humane mission to try and wean the Spanish off bullfighting and onto rugby. I do recall I found it a bit disconcerting that their wing forward kept waving a cloak at me and sticking a kebab skewer in my neck as I ran past him. (No, not really. Joke...sort of.)

It was actually a jolly enjoyable tour, notable for at least four vivid, yet oddly random, memories. I remember the lunches: coss lettuce with olive oil on it, served with a crispy bread roll. (I hadn't been abroad before, so this was wonderfully exotic!) I remember the romance: I fell hopelessly in love with the daughter of a local dignitary. She couldn't speak a word of English but boy could she twist! ('The twist' was a dance craze of the time which involved no bodily contact whatsoever between partners. Pity.) I also remember the rugby. Especially the climactic game in Barcelona, where our college team beat Spain. Yes, the national side. (It wouldn't happen now, of course; didn't they just qualify for the World Cup?)

Above all, though — of course — I remember the birds. Seeking to celebrate our victory in a slightly more edifying manner than reciting Eskimo Nell with a pint of beer on my head, I set off in search of some feathered friends. Thus, I found myself in Barcelona docks looking down on a few straggly bushes along a stone wall by the quayside. And it was there, in this least likely of 'sites', that I received my due reward for converting the winning try. My notes from that afternoon are brief yet satisfying: "Many migrants in: Great Grey Shrike, Hoopoe, Subalpine, Dartford, Orphean, Sardinian and Bonelli's Warblers. Rock Bunting and...an Alpine Accentor."

Little did I think then that it might be the only one I'd ever see! Unless...OK. It's now the morning of 15th. Call Birdline: "Alpine Accentor still showing well at Rimac." Shall I? Did I? No, I went into the garden and practised my place-kicking.

Holiday blues
Morocco at its worst...and best

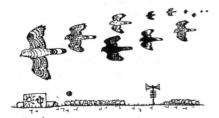

I normally manage to eke some birdy satisfaction out of a family
holiday, but in August 1991 I reckoned I might finally have
failed. We were staying at a Club Med in Morocco, but not on
the west coast near any of those marshy Oueds — haunts of
Bald Ibis and Slender-billed Curlews — nor within day-trip
distance of the desert with its wheatears and sandgrouse. Nor
indeed was it anywhere near the delightful Atlas Mountains,
home to Alpine Accentors and Tristram's Warblers.

Oh no. We were on the north coast, where the bushes have
been devoured by goats and the wetlands obliterated by
developments, where any fertile land has been cropped and
sprayed, and where everywhere else looks like an unofficial
refuse tip. Morocco might just be one of the best birding
countries in the world but Smir (pronounced 'smear') might just
be the only birdless bit of it. That was where the Club Med was.

And was it fun? Well, we referred to our rooms as "the John
Macarthy Suite", and survival there as "taking a Smir Test". OK,
hostage and gynaecology quips may be in bad taste, but then so
was just about everything about that place. My only respite from
the agony of being mimed at by French drag artists, kept awake
all night by drunks, throwing up after the 'authentic local feast',
and having my favourite T-shirt stolen off our washing line, was
to look forward to escaping for a day's birding.

Not that I was very optimistic about it. On my few foot-bound
explorations outside the perimeter fence so far, I had seen very
few signs of feathered life, apart from some scruffy Cattle Egrets
poking around in a pile of festering filth, under which was the
remains of the only local site mentioned in my *Where to Watch
Birds in Morocco* book. The signs weren't encouraging and, as I
set out, I was already feeling resentful at the exorbitant cost of
the car hire. I had no more specific plan than to head for the

narrowest part of the Straits of Gibraltar. I wasn't sure what would happen when I got there. Maybe I would attempt to swim to Spain; or just leap off the cliffs and end it all; or — if I was lucky — I might see a few migrating raptors.

So what did happen? Well, I've just been re-reading my notebook account of that day. Here's what I wrote at the time: "August 31st. The big day out! I was somewhat frustrated by the fact that I couldn't get the car till after 9 am, but was consoled by my belief that 'little birds' seem more or less non-existent round here, so I probably wasn't missing anything! Anyway, I set off through the nearby town past horrible camp sites and ghastly rubbish, and eventually the road began to climb among the 'mountains' by the Straits.

In no time at all, I found myself under a swirl of Honey Buzzards and Black Kites, with a Booted Eagle in for good measure. It was immediately obvious that birds were all over the place. The map bore little relationship to the roads, and whether or not I actually found Punta Ceres — the narrowest crossing — I really don't know. It didn't matter. I simply parked the car at a watchpoint on a hill rise and enjoyed the birds streaming past from the north. They obviously cross the water at about hilltop height, so the lowest can be really close: even eye contact was possible. Then they find the thermals, swirl up and move off south in the air streams."

I watched and counted from 10.30–12.30. I soon got to appreciate the approach and altitudes of the various species: "Short-toed Eagles fly lowest, usually in ones and twos. The Booteds pass by just above them in twos and threes, once with eight together, often pausing to circle around before carrying on. Egyptian Vultures fly above them, and then the Black Kites, except the ones caught up with the real high-flyers — the Honey Buzzards — which occur in the biggest flocks. It's as if the whole thing is regulated by air traffic control!" So in two hours the log comprised: Honey Buzzard 600, Black Kite 30, Booted Eagle 48, Short-toed Eagle 17, Common Buzzard two, Montagu's Harrier three, Marsh Harrier two, Egyptian Vulture 10, Sparrowhawk two, Kestrel, Hobby and Peregrine one each.

There were more pale than dark Booted Eagles, some very ghostly Short-toeds, and as for the Honey Buzzards, well, what a range! I quote again: "The dark-coverted juveniles are superb, and the adults are amazingly variable with just a few having almost white heads. Tail bars are 'classic' on some, while others get pretty indistinct. But any close views show some degree of horizontal underwing bars. Lovely."

I then took a detour inland into the mountains in search of 'rock' species. Total failure. The habitat was very bleak and looked fine for Black Wheatear, Rock Bunting and Rock Sparrow — but I saw zilch. I constantly, however, came upon raptor swirls. Then, as I began to descend again I decided to do another count from a point opposite a very narrow part of the Strait. It was spectacular indeed. Although birds were a little higher by now, the Honey Buzzard invasion was truly breathtaking. At 14.10 I had over 1,000 Honeys in sight above me, stretching back in ranks as far as the eye could see, across the water to Spain. During the next half hour I saw at least as many again: over 2,000 Honey Buzzards in half an hour, with a few Black Kites, 'Gippos' and Booted Eagles. As wonderful a migration as I've ever seen. In contrast, the nearby woods produced two Chaffinches and a Chiffchaff!

That evening I told my family what a great day I'd had, while being 'entertained' by a six-foot Frenchman wearing leather suspenders and a pair of plastic bosoms. Ah yes, how much more we appreciate our pleasures when we've had to suffer for them.

Holiday special
Portugal

First mornings — I love 'em. Latest example: 28 August, eastern Algarve, southern Portugal. Never been there before, family holiday, but the Air Portugal in-flight magazine had already promised me that it would not be birdless. I quote: 'Among the species that nidify there and live throughout the year or stay there in between their migration flights, the sultan bird detaches, an extinguishing species. Pink flamingos, wrynecks, shovellers, snowy egrets, rabbit hunters and storks live in the lakes and in the surrounding fields, over which discreet observatories enable us to follow their habits.'

Boy was I looking forward to this! I'd never seen a sultan bird (detached or otherwise) nor a rabbit hunter (assuming it was a bird); let alone the first proven European record of Snowy Egret. And a discreet Observatory? Well, as it turned out, that's exactly what our first night accommodation resembled. Due to a mix up with the villas we had to sleep in a glorified concrete hut, with a rusty stove and hot and cold running spiders. My wife and daughter weren't too impressed, but to me it felt like the bridal suite at Cape Clear. I loved it.

Things got even better when I tiptoed out before first light and found the air filled with wader calls. Whimbrel, Grey Plover, Greenshank. And what d'ya know? — we were staying 100 yards from the edge of a tidal lagoon. What a bit of luck (tee-hee). It was barely light as I padded down to the coastal road (well, sandy track actually). There I immediately encountered the inevitable drawbacks of any Mediterranean dawn experience: dogs, gunshots and pigeons. Portuguese pooches (like their Spanish or Italian counterparts) seem incapable of hearing a distant footstep without scampering over to attack it in a fit of snapping and snarling. Mind you, you only have to wave a toe in their direction and they fall over on their backs and whimper pathetically. (I could draw a parallel with certain footballers from the same region, but I won't.) The gunshots I found rather more worrying, because I assumed the bullets would be aimed at birds, and also the pigeons soon began to annoy me because I immediately began to misidentify them. Within five minutes, I'd strung a White-winged Black Pigeon over the marsh, and a flock of Black-bellied Sandpigeons over the horizon. Then I calmed down and accepted that at least 99.8% of the birds I was seeing in the still half-light were domestic doves, and I'd better get used to it. Happily, it was at this moment that I also realised that the 'gunshots' were merely explosions from some kind of bird-scaring device at the local farm and, what's more, it seemed likely that the birds the farmer was most intent in keeping off his crops were in fact the pigeons. Nice one. My anxieties dissolved, and I turned my attentions to the nearby estuary. Within the next hour, I ticked off no fewer than 21 species of wader, including Kentish Plovers and Black-winged Stilts. It was like having Titchwell to myself!

Which was all very well, but this was Portugal. So surely I should expect something more exotic? I soon got it. As the sun rose and glowed more brightly so did the birds. Two Hoopoes flopped across the path; 60 Bee eaters blooped over my head; a gaggle of Azure-winged Magpies chattered through a nearby orange grove; and a large green parrot squawked through the eucalyptus trees. A parrot? Yup, there's quite a few of them flitting around the Algarve, no doubt testimonies to the incompetence of Portuguese cage-makers. But what about the flocks of Waxbills? Are they escapes gone feral, or are they spreading from Africa? (I saw the same phenomenon in Trinidad last year. Is Waxbill due to do a Collared Dove?)

All this variety, intrigue and colour! You'd think I would have been ecstatic. Well, I was enjoying it, but — call me perverse and ungrateful if you like — there was one element I was still craving: little brown migrants. OK, our hut looked like a bird observatory, so where was the 'obs garden'? In fact, I'd walked straight past it when I'd left before sunrise. Now it was nicely lit, and it looked rather promising. It was obviously an ex-allotment, now mainly neglected. There were patches of bramble, and fennel gone to seed and, best of all, a large fig tree covered in half-rotten fruit. Previous Mediterranean experience has taught me that birds seek out such trees; and therefore so do I. Here was my very own. I sat about 10 yards away and scanned through the shiny leaves and shady branches. A good fig tree rule: take your time. There's always more in them than appears at first glance; and also birds come and go. And did they? Well, during the next half hour, my tree produced: one Pied and two Spotted Flycatchers, adult and juvenile Woodchat and Great Grey Shrikes, a Nightingale, three or four 'Willowchiffs', two Garden Warblers and two Whitethroats, Melodious and Olivaceous Warblers side by side for comparison and, my personal favourite, a subtle little Bonelli's Warbler that eventually came so close to me that I couldn't focus down on it (and using Leica 8 x 32s that is very close!). Not a bad selection, eh? And all the more entertaining because it just could have occurred in a British bird observatory garden, on a very good day! Thus do my little fantasies keep me happy.

And thus did I return to my family and declare my first morning in the Algarve a little triumph. I'd like to tell you that the garden continued to produce migrants for the rest of the week, but the truth is it didn't. But it might have done if there'd been a rainstorm or two. But then the fact that it was continuous blue skies meant we did have a superb holiday, especially after transferring into the splendid villa up the hill. If you wish to sample its delights, it's advertised in RSPB Birds magazine. I shall say no more. Except that if you do try it, and you grip me off with Sultan Bird or Rabbit Hunter — I'd love to know what they are!

A shaggy dog story
Taking the woof with the smooth

Yappy dogs scare me. Call me a wimp if you like, but I just don't enjoy being barked, growled, or snarled at. It happens a lot on my local patch, Hampstead Heath. Dog owners come up with some incredibly naff excuses on behalf of their belligerent pooches, the most surreal I've had being: 'He obviously doesn't like your aura!' To which the obvious reply is: 'Madam, I cannot control my aura, any more than it seems you can control your dog.' Well, that's what Oscar Wilde would have said. Not that he went birding on the Heath (though he would no doubt have found plenty to his taste up there on a Saturday night). But I digress.

Anyway, for me, one of the joys of birding on holiday — i.e. not on the Heath — is hopefully no dogs. Imagine then my consternation when I was in the Algarve in late August and set off for my first early morning bird walk, only to find myself accosted by not one but two of the local hounds. Dog number one looked like a cross between a Jack Russell and a Corgi, and he was at least friendly — to a fault. He leapt at me with mud-covered paws and licked my shorts (been watching too much Bart Simpson perhaps). Dog number two was less chummy. Bigger, mongrelly and black, he was clearly a Doberman wannabe. He bared his teeth and took the Bart Simpson instruction more literally by attempting not only to eat my shorts, but also by nipping at my shirt tail. I suppose I should commend his accuracy for not grabbing a chunk of me as well. Instead, both dogs grabbed a piece of each other. They started rolling around and snarling and snapping. I attempted to walk on. But they came with me, continuing their fight — between my legs!

I tried everything to discourage them: 'No', 'Stay', 'Vamos', climaxing with a full-throttle, eye-bulging scream of 'f*** off!' But they didn't get the message. I trudged towards the saltmarsh with both of them attached to my ankles as if by elastic. I really was feeling very seriously stressed; but perhaps escape was nigh. Immediately ahead was quite a wide and possibly deep tidal channel. I needed to cross it to get at the promising looking scrub and bushes on the other side. Maybe this is where I would lose both dogs. Wrong! Number one immediately leapt into the water and doggy-paddled — is that the only stroke they can do? — across to the other side, where he stood on the bank panting and challenging us to take the plunge. However, Doberman Wannabe was not so bold. He dipped a paw in the stream and then started whimpering pathetically. His cowardice gave me strength. Fearlessly, I strode across. In fact, the water barely came up to my knees, but no way was Doberman Wannabe going to follow. He panicked up and down, yowling the doggy equivalent of 'Please don't leave me'. It was at this point, at a safe distance, that I sarcastically named him 'Fang'.

So, one down, but definitely still one to go. However, he was obviously going to go with me. I was just going to have to accept him as my birding companion. I named his Carlos (about the only Portuguese name I could think of). With the sibilance of the local accent this became Carlosh, and that was rapidly corrupted to Carwash, because it's the title of a favourite Seventies groovy movie, and it was also what he needed to go through now that he was covered in black oozy mud.

Although I was resigned to Carwash's company, I wasn't entirely glad of it. His main joy seemed to be bird-scaring. Ringed Plovers, Stilts, Whimbrel, Turnstone, even Fan-tailed Warblers, he leapt at, chased, and attempted to snap in his jaws. Then it dawned on me: maybe he was just 'doing his job'. Maybe he was a hunting dog, used to flushing up birds to be shot at. I did notice that he was rather efficient: snuffling ahead of me, round bushes, under hedges. What's more, when I stopped, so did he. I decided to be positive. I appointed him Carwash the bird dog, and I started talking to him (barely less screwy than my usual habit of talking to myself).

'OK', I told him, 'We're going birding, not hunting.' I could hear something 'takking'. 'What's that calling?', I asked Carwash. He rummaged in a bramble and out shot a Whitethroat. 'OK, now try that hedge'. In went Carwash, out came a Pied Fly and a Woodchat. We arrived at an orchard. 'Privado. Prohibido Entrada', said the sign. I always obey Mediterranean signs, mainly due to my fear of Alsatians and men with guns, but Carwash was in there like a flash, rooting out several 'Phylloscs' and a couple of Nightingales. He was obviously a very bright dog.

So I decided to really test him: 'Go seek Olivaceous Warbler.' Carwash romped straight up to a fruiting fig tree. Out popped a Garden Warbler, two Sardinians and…a Melodious Warbler. OK it was a rather dowdy one, but a Melodious, definitely. Just as I thought, dogs may be intelligent, but they are crap at ID. If he'd done that on Scilly, he would have been labelled a stringer for life. I pointed out the features: 'see the faint lemony wash on the throat' — and he looked suitably shame-faced. But Carwash was a fast learner. On the way back, he found two more Melodious (his apologetic shrug told me he knew what they were) and — just to show he wasn't rarity-obsessed — he ran off into a vine field thus leading my gaze to a Little Owl perched on a stump and sunning itself. Nice one, Carwash.

Back at the villa, Fang was waiting to greet us. No more shirt-tugging or snarling. Just panting admiration for Carwash. I too had been impressed. If we'd been in Morocco he would have demanded money, pens or cigarettes. Instead, all he asked for was an appreciative pat. I gave him one, and he licked my hands (so much tastier than my shorts). As I munched breakfast figs (courtesy of Carwash) he lay snoozing and twitching in his sleep. Dreaming of chasing Olivaceous Warblers, I imagine. 'Mmm', I thought, now that the quarantine laws have been relaxed…I wonder how he'd get on at Hampstead Heath. I'd settle for a Melodious there.'

Have I got smews for you!
Holland in winter

It seems like almost every day someone asks me: 'What's your favourite duck?' (Well, someone did once — in a Chinese restaurant, actually — but I've got to start this piece somehow.) The answer is: male Smew. Right little crackers aren't they? They look as though they were originally fashioned from pure white porcelain, then dropped and smashed, but then glued together again, leaving delightfully zigzaggy cracks, and with only a black eye to remind them of their mishap. And not only do they look great, they've got a splendidly silly name. Where on earth did 'Smew' come from? I looked it up in the endlessly entertaining *Oxford Dictionary of Bird Names*. 'Smew: a Norfolk word, also recorded as smee or smee duck, which imitates the whistling sound heard from this species. Also denotes Widgeon. The Dutch word smient and the German schmiente also mean Widgeon.' In other words, they misidentified it. Typical of the twerps who named our birds: couldn't even tell a Wigeon from a Smew. Never mind, male Smew is still my favourite duck.

The problem is, of course, that male Smews are not terribly easy to see in Britain. I birded the Birmingham reservoirs for some 15 years, and saw only two or three red-heads (which, from now on, I shall call brown-heads, in case any of the twitchers get upset at the idea that the Redhead they ticked off last year wasn't actually a British first — (which of course it wasn't. It was an escape; but I digress). In all my years of Midlands birding, I saw only one male Smew. Mind you, that was a pretty unforgettable apparition. It occurred sometime during the legendary bleak winter of 1963, which covered the supposedly unfreezable waters of Bartley Reservoir with so much ice that there was only one small duck-sized patch left: a space occupied one memorable morning by a male Smew — in a snow storm!

It was a vision that gradually faded over the next phase of my life — the rest of it — during which time I have lived in London.

There was a certain irony in the fact that, for a long time, my rather sporadically-visited local patch was Brent Reservoir, which, just after the War, was known as the Smew Mecca of southern England. Indeed, I was originally inspired to go there by pictures of Smew flotillas in ancient books. The photos were in black and white, but that was OK 'cos so were the birds. Sadly, by the time I got to Brent, they were long gone and, in all my visits, I saw no more than — you guessed it — two or three brown-heads.

This last winter there was, of course, something of Smew invasion to southern England. Indeed, on 25 December, a single bird — brown-headed, naturally — even had the complete lack of judgement to pitch down on the only unfrozen puddle left on Hampstead Heath. It was no doubt thinking: 'What a way to spend Christmas!' But it was a nice present for me, and I got the closest views of the species that I have ever had outside of a Wildfowl and Wetlands Trust Reserve. Nevertheless, call me ungrateful if you like, its very closeness merely served to remind me that it wasn't the porcelain beauty I so craved.

Which brings me to 10 March 1997. We are in Flevoland, Holland, enjoying a *Dutch Birding* weekend. 'We' are Derek Moore (Director of the Suffolk Wildlife Trust), Richard (never did discover his surname, from English Nature), Chris Packham (TV Presenter, ace photographer, all round naturalist and avid birder), and me. Our car edges out along what appears to be a road to nowhere, across a seemingly endless ocean. In fact, it is a causeway, crossing a vast man-made lake; part of the astonishing reclamation that went on in Holland barely 20-odd years ago. We pass a warning sigh: 'VOGELS!' This is Dutch for 'birds'. We look out of our windows, and there they are: vogels unlimited. Scores of Great Crested Grebes, stacks of Goosanders, thousands of Tufted Ducks and...what are those little flocks of white things? Gulls? Up with the bins. No, they're Smews. Pack upon pack of them. There are brown-heads, yes, but real sexy little females, rather than the dozy juveniles we mainly get in Britain, and each one is surrounded by a posturing posse of gleaming males. This is what Derek had promised to show Chris. Chris was duly impressed: 'Stonking. Top Ten. Serious Smewgasm.' For an amalgam of knowledge, enthusiasm and hip linguistic eloquence, Packham is a hard act to follow! I too was impressed. Especially when the males starting displaying, dipping their chests, and flourishing their shaggy white hairdos. Then it struck me: I'd seen this look before.

Remember Chris Packham in his "Really Wild" days? Peroxided and spiky. He'd told us that he was going through his punk phase; his coiffure presumably modelled on Billy Idol, or a Sex Pistol. But clearly not so. Maybe even Chris wasn't aware of the now incontrovertible visual evidence swimming before us: what he, in fact, had really wanted to be was not a punk icon, but a male Smew!

Since that day, an irresistible train of thought has increasingly obsessed me. Goosander, Mergansers, Smew: they all have the punky hairstyle, right? I bet Chris still has his guitar in his attic (and I bet he hasn't forgotten those two-chord riffs). I've still got my drums. Derek Moore used to be the lead singer in a rhythm and blues band (not a lot of people know that), and Richard...well, he can play the bass, 'cos even if he can't, no one can ever hear it. So if the old TV presenting and Conservation gigs begin to dry up, how about it lads? 'Hey, hey, we're the Sawbills.' Sounds good to me. Dream on.

Owl quest

I've started, so I'll Finnish

In mid-May, I went to Helsinki. I was there to take part in 'The Battle of the Towers': a misleadingly belligerent title for what turned out to be a delightfully amiable event. The towers in question were not as in Blackpool or Eiffel, but as in two-tier open-top bird hides, from which Finnish birders are in the habit of scanning across marshes, woodlands and, in particular, the skies, spotting migrating birds. On 17 May, several teams climbed up wooden steps in various parts of Scandinavia, not just Finland, and did exactly that. The aim was not merely to count and compete, but to draw the public's attention to what an absorbing and enjoyable hobby birding is. I confess there was a moment when I wasn't entirely convinced that identifying an extremely tiny dot as a sub-adult Lesser Spotted Eagle rather than a wasp was really the way to turn on the dudes; but in fact there was a constant stream of curious and enthusiastic visitors to our particular tower, and a splendid time was had by all. The whole thing was organised by the doyen of Finnish birding, Lasse Laine, whose fame, expertise and energy is acknowledged right across the birding world. I was Lasse's guest.

Now, much as I enjoy a good tower battle, I was also anxious to fill in a few gaps in my European list during the couple of days I had either side of the event. Thus, when Lasse met me at the airport, I immediately enquired about a number of Finnish specialities. Unfortunately, however, though I'd browsed through my Lars Jonsson, I hadn't bothered to look at a map, or indeed take account of the date. Suffice it to say that some of my requests were the equivalent of a 'foreigner' landing at Heathrow in the middle of February and asking to be taken to the nearest breeding Ospreys. So Lasse told me first the bad news: it was too far for Siberian Jay and too early for River Warbler. But then the good news: four out of my six desirable Finnish owls were available within a few hours of Helsinki. And boy did Lasse deliver. I warn you, if you are of a jealous disposition, what you are about to read may get you just a teenie bit envious. Otherwise, share my excitement.

First up: Eagle Owl. It was actually mid-morning as we scoped a cliff face at least half a mile away until we finally focused on what looked like a couple of fluffy teddy bears. Eagle Owl chicks, on their nesting ledge. Very sweet, but rather dozy, and far too cuddly looking to truly live up to their fearsome name. I was happy to tick them, but somehow it didn't seem quite right. Nevertheless, surely mum and dad would be snoozing invisibly somewhere in the deep dark forest. Or were they watching us? We walked on past the cliff and approached from behind, along the ridge, as it were. Presumably Eagle Owl parents like to check out anyone who attempts to visit their children, because suddenly there was one of them. Male of female? Who cared? It was huge. Perched in a pine tree, bigger than me; the owl I mean, not the tree. Orange eyes blazed at us as I flapped around with my tripod. Then it also flapped, on impossibly massive wings. But not away. It came closer, landed, posed again, and glared into my zoom lens. And this time the eyes filled the frame. Then I suppose it recognized Lasse as its friend, and me as a harmless admirer and finally flew, till it dissolved into the fir trees, disappearing as mysteriously as it had materialised in the first place. One down. Next, please.

A different part of a different forest. Less open this time. Murky and silent. Not even the expected small birds called. How come Finland is wall to wall pine trees yet Coal Tit is rarity here? Or maybe they knew that this is the domain of Ural Owl. This one was a nestbox job. Except you don't go near a Ural Owl's nest box unless you have a death wish or a crash helmet. As we tiptoed forward, Lasse informed me that people have lost hair and eyes to protective Ural Owl parents. My *Birdwatch* baseball cap seemed strangely inadequate. I became genuinely nervous. And even more so when Lasse's gesture directed my gaze half-way up a pine tree, barely 20 yards ahead of us. And there it was. The dangerous demon of the forest. Rather like an enormous grey Tawny Owl, but with dark doleful eyes looking so sad that you wanted to shin up the trunk and give it a big hug. I resisted. Instead, I ticked it. OK. Two down, two more to go.

From the huge and dozy, to the tiny and hyperactive. Pygmy Owl compensates for its diminutive stature by being absurdly energetic. I know that syndrome precisely. Its nestbox was barely a metre off the ground. When Lasse scratched gently on the side, the bird shot out like a stone from a catapult. Can you tick a blur? No need. The little beauty landed on a nearby branch, where — to use that usually rather silly birder's term — it proceeded to 'perform'. Except that in this case the verb was entirely appropriate. It cocked its tail like a Wren, waved it around, leaned to both sides, spun its head almost through 180 degrees and, just to emphasis how small it really was, flitted onto another even closer branch in order to provoke two Tree Pipits into mobbing it. More fool they, since despite being barely four inches tall Pygmy Owls can kill and carry off birds almost twice their weight. But this one wasn't hungry. It was just showing off.

Which was more than could be said of the fourth and final member of the Owl quartet: Tengmalm's. This time the box was at least 30 feet up a lone birch tree, as branchless as a telegraph pole. Lasse announced our arrival by a single tap on the base of the trunk. Whereupon, the owl stuck its head out of the hole and looked down at us, with a bemused expression in big round eyes that could have earned it a puppet series on children's television. Talk about cute! For five minutes it stayed there, barely blinking, until its eyelids began to close and, as if muttering 'Can I go in now?', it almost slid back down into its nest.

So that was it. Two days, four new owls. All of them different characters. All of them totally unforgettable. Envious? I would be. Grateful? You bet I was.

By the way, if you fancy a Finnish Owl quest, please don't bother Lasse, he's a very busy man indeed. But there are other local guides who have the necessary knowledge if you go on an organised trip. And if you have time to go up to Lapland, you may get Hawk and Great Grey Owls as well. Oh, and one final thought for this country: How about lots more owl nestbox schemes. They really do work. Ask the owls.

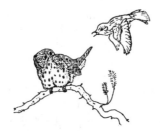

The American way
You never forget the first time

When the spring migrants pass through the United States on their way north, American birders call it the 'warbler wave'. It's not like the Mexican wave, though come to think of it the birds are hardly less ostentatious.

Yankee warblers are right little exhibitionists. They travel in packs, they flit around a lot, and they are quite incredibly colourful. They wear coats patterned with spots, streaks and stripes so gaudy that they make the star-spangled banner look like an anorak. I've suspected that they have to pass a 'razzy test' before being allowed to enter the country. Our dreary little British warblers would never make it through immigration. (Be honest, most of them do wear anoraks!)

I'm tempted to muse that American warblers have to be so conspicuous to make it easy for American birdwatchers to identify them. American birders are not, by and large, known for their patience with 'difficult' groups. I recall, on my first trip to the USA, meeting an Audubon Society field meeting at Jamaica Bay who refused even to attempt to help me sort out the splendid variety of waders on offer: "Oh, we don't bother with peeps...too tough for us". As it happens, the same group then asked me to help them sort out the warblers! In fact, it wasn't too difficult that day, since it was early November and the only ones present were several hundred Yellow-rumpeds.

But it's a different matter in spring. I'll never forget the first time I encountered the warbler wave. The circumstances were somewhat bizarre. In 1964 and 1965 I had been working in the States, along with John Cleese, Tim Brooke-Taylor and others, first performing on and off Broadway, and then touring with a stage version of what was billed as David Frost's That Was The

Week That Was (a production most notable for the fact that David Frost wasn't actually in it; which almost got us lynched in several cities. They should have been grateful!)

Anyway, in early May we were back in New York, gathering up our belongings, ready to return home. On the evening of 3 May I went for a walk in Central Park. Most New Yorkers wouldn't do this, as the park is the crepuscular haunt of all sorts of dodgy life-threatening characters, from muggers and drug-pushers to mime artists. But British birders know no fear. As darkness fell, I became aware that little shapes were flitting around the tree tops. The spring migrants were arriving. I felt a frisson of excitement. However, there were two problems; first, it was getting dark, and second, I was flying back to London at 10 o'clock the next morning.

So it was that, at 5 am on 4 May 1965, I found myself perched on a rock in the middle of The Ramble in Central Park, New York City. It was barely light, but the sky was alive with squeaks and tsips. Then, out of the sunrise cascaded showers of small birds. Suddenly, the branches around me were — as birders so unromantically put it 'heaving' with warblers. All so colourful, all so conspicuous, obviously several species. But which was which?

At first I panicked. There were so many, and they wouldn't keep still! Then I started taking notes. Spots, streaks, patches of yellow, flashes of white. Where exactly were the markings? Coverts, scapulars, rumps. I tried making thumb-nail sketches. That's what they looked like — thumb-nails! Each time I looked up, the bird had been replaced by something else. This wasn't working at all.

Then I remembered American logic. Several of their birds' names are utterly literal. Describe the bird and you've got it. Black-and-white Warbler. Fine, that's that one. Black-throated Blue. Black-throated Green. OK. But how about Blackburnian? Does that one look like Alan Shearer? And Worm-eating: does that lose its identity if it drops the worm? And Magnolia? Only in magnolia trees? I don't think so. That's a magnolia tree and there are at least three species in it. Even Yellow-rumpeds weren't so simple: back in the Sixties they were still called Myrtle Warblers, and so much more charming for it, I'd say.

So, was my first warbler wave frustrating? Not a bit of it. It was one of the most thrilling birdy experiences I've ever enjoyed. I was only there for a little over an hour, but after the initial confusion I realised I simply had to watch and enjoy. Memory did the rest. On the plane back to England I gazed at the American field guide and ticked off what I knew I'd seen. I may have missed one or two, but I am absolutely certain that in that hour I'd recorded 17 species of American warbler; not to mention various vireos, tanagers, orioles and thrushes.

I've witnessed American warbler waves a few times since. Some of them have been less impressive — they say that's generally been so in recent years — though some of them have involved even more species, even in Central Park. But, just like that first sexual experience, you never forget the first time. And which is best? Well, you tell me.

A tale of two Godwits

Everyone's a winner...it just depends where you are

One of the ironies of birding abroad is that species that are 10-a-penny in Britain may well be great rarities wherever you happen to be. Mind you, it's not always easy to share local enthusiasm for a bird that you didn't even realise was rare in the first place. For example, I remember passing on my wader count from Mahé Harbour on the Seychelles. Lesser and Greater Sandplovers, Kps and Terek Sandpipers got merely a shrug, but when I mentioned two Oystercatchers the recorder gasped: 'What? They're first for the Seychelles!' But it didn't really make them any more exciting for me. I was similarly unmoved by a second-year Lesser Blackback on Trinidad — only about a dozen records. But that lack of emotion paled besides the total apathy I displayed in Papua New Guinea, where the highlight of a day watching Rainbow Bee-eaters and Birds of Paradise in the rainforest was, according to my local guide, the discovery of a European Starling at Port Moresby sewage farm.

This British bird abroad syndrome is most rife in the United States, where complete duffers like Black-headed Gull and Tufted Duck can spark major twitches, and the report of a Blue Tit would probably cause national gridlock. I experienced the phenomenon myself when I was filming on Martha's Vineyard in Massachusetts (near Cape Cod, Boston, that sort of area). It was August, less then 10 years ago, but the ambience was pure Seventies. One afternoon, we arrived at a wooden-slatted mansion that looked as if it had been built specially to be blown away in the film "Twister". Fortunately, no such dramas occurred that day, although it has to be said that many of the people in the house were already — in a manner of speaking — flying. Around, on and under a huge wooden table was an assembly that looked as though they had been gathered together for a Grateful Dead reunion rockumentary. Lots of long hair, collarless shirts, cheese-cloth pants, funny cigarettes and a bowlful of something white that I'm sure wasn't sugar. Rest assured that I immediately took refuge in my inhibited Englishman mode, although to be honest I felt instantly at home. I also felt about 20 years younger. It was incredibly friendly, but also unbearably noisy. The Best of Little Feat blared out at a volume which totally precluded conversation, but that didn't stop everyone talking. Well, I say talking...in fact, the only intelligible sounds I could decipher above the babble were yelps of 'Wow!', 'Unbelievable!', 'Too much!', 'Al-right!', and the like; and I swear I even heard the odd 'far out' and 'groovy'. Our host was called Earl, or Duane, or Gregg — or has my memory rechristened him to fit the image? Well, he certainly wasn't a Humphrey or a Julian. Oh yes, I remember, he was called Vernon. Or rather, Verne. Or rather, 'Big Verne' ("but you can call me 'Big'"). Physically, he was a sort of Meatloaf x ZZ Top hybrid. Glibly one might have called Verne, his domain and his entourage 'hippy', but something about the extravagant chaos also suggested considerable wealth (that stuff doesn't come cheap you know). In fact, Verne was also a real estate king (that's a property dealer in English), and on Martha's Vineyard — where many a movie star has a holiday home — that means big bucks. More importantly, Verne was also a birder, and so were several other people in the room. I began to realise this as certain birding-type Americanisms detached themselves from the hubbub: 'Let's hit the high-tide roost.' 'We'll nail that mother.' 'Go for it!' 'Yeah. Right!' 'Wow! Unbelievable!' If I had interpreted their hyperactive enthusiasm correctly, they were planning a twitch, to somewhere called Monomoy Island. 'Monomoy? Is that good?', I enquired. 'It's unbelievable. Wow!' I plucked up courage: 'Can I come with you?' 'Sure. Wow! Unbelievable. We'll charter a plane.' 'Wow! Unbelievable' (that time it was me). 'Oh by the way', I asked, 'what bird are we after?' 'Bar-tailed Godwit. Wow! Unbelievable.' Oh.

It was no ordinary twitch. Early next morning about 10 of us crammed into a light aircraft — you know, the kind that disappear mysteriously over the jungle; or maybe over off-shore islands near Cape Cod. The pilot did his best to follow the storyline. As we flew over a sand bar he suddenly yelled 'OK guys, let's hit the deck!', at which the plane plummeted almost vertically downwards, did the aerial equivalent of a skid turn, and hurtled along for half a mile with the undercarriage almost skimming the waves. The American yelled 'Wow! Unbelievable'. I nearly threw up. A few minutes later, we landed at Chatham, on the Cape. There, we were met by a (or do I mean the?) quiet America, Middle-aged, portly, check-shirted, baseball-capped, called Al or Mort, or something similarly comforting. We all squeezed into Al or Mort's pick-up, and trundled down to a jetty, where we tumbled out of the truck, and into a speedboat. We tazzed across the channel (even Als and Morts can't resist the thrill of an outboard and flying spray) and landed on the sandy beach of Monomoy Island. Out of the boat. Deep breath. Quick chorus of 'Wow! Unbelievable', and off we galloped to 'nail the mother'. Or to 'try and find the Bar-tail', as we Brits would say, at a much reduced volume.

Three hours later, I was utterly absorbed. I had sorted through the dowitchers: 'juvenile Short-billed have the internal tertial bars don't they?' 'They're brighter. They're easy', commented Verne. (So much for feather waffle.) I'd picked out a White-rumped Sand. 'Good call, Bill!', (Verne was just being kind. I heard the mutter of 'trash bird'). And I'd even tried a little British one-upmanship. 'OK, let's see if I can find a Little Stint', I announced, as I scanned the hundreds of Semi Ps. But Verne dismissed my attempt instantly: 'They're impossible. You can't call 'em.' (Mmm, I thought, tell that to Anthony McGeeham.) Never mind, I was having great fun. But as time passed, I realised my transatlantic chums were getting quieter. Did I sense American enthusiasm waning? Did my ears deceive me, or was this the sound of silence?' Then I remembered what we were there for. Not the hundreds of Yellowlegs, nor the Semi's (plovers and sands), nor the Forster's Terns, alongside Roseates and Leasts, nor even the breath-taking spectacle of a roosts of 2,000 Black-bellied Plovers (Grey in the UK, but these mostly did have black bellies). No, forget all that, we were after Bar-tailed Godwit. Just one. A first for Massachusetts. But it was nowhere to be seen. Verne and his mates had dipped. I apologised (it was probably a British bird, so I felt personally responsible). I was very sympathetic. On the other hand, I was rather enjoying the sudden lack of loud voices. I was also enjoying a new bird. Barely 30 feet away, driven up by the rising tide, nearly a hundred of them, clustered together in a feeding frenzy, the species that I had secretly hoped to see on Monomoy. 'Oh Yes! Look at those!' My companions turned to see what I was enthusing about. I couldn't resist it. 'Godwit! Hudsonian! Wow! Unbelievable.' Verne laughed. You see, Americans do have a sense of irony.

Let's pish again...
Like we did last autumn

Early in November I spent a few days at Cape May, New Jersey, USA. There, I bought a splendid little book that is essential for all visitors: The Birds of Cape May by David Sibley. It begins with a section on birding etiquette which includes the sentence: 'Please keep pishing to a minimum. It doesn't work very well anyway'. Well, I don't wish to sound as if I've been smitten by the spirit of the panto season, but my response to that is: 'Oh yes it does!' Well, it does in America anyway.

I first realised this about 10 years ago, when I went wandering in the woods with Pete, a birder from Boston. Believe me, there are few more bewilderingly birdless places than a New England woodland in early fall. It is all the more frustrating because it looks so damned good. The trees are massive, the canopy is wondrously verdant, and the effect is like standing inside some vast lofty natural cathedral — an illusion compounded by the almost eerie echoey silence. Until Pete started pishing. And when I say 'pishing' I mean pishing. I don't mean any of that namby-pamby effeminate kissing-the-back-of-your-hand stuff. (Try making noises like that in Massachusetts, and you'll either pull one of the Village People, or get lynched.) I mean good ol' tongue-tingling, lip-trembling, all-American pishing, just like momma taught you. In case you still don't get the idea, you put your lips together and go: 'Pish! Pish! PISH!', loud and long. In fact, Pete went on for 10 minutes with nothing happening, except his face going red with hyperventilation, and mine red with embarrassment. 'It's OK', I said, trying to let him off the hook (he'd promised me some Yankee warblers), 'There's obviously nothing here.' But Pete simply pished again. 'Keep at it, and something will come', he assured me. And it did. I first spotted a flitting movement about 20 yards away, but I suspect the bird had been summoned from nearer half a mile. An immature Chestnut-sided Warbler. Just one bird, but it was a tick for me, and — let's face it — it had materialised in a place that had seemed as lifeless as a graveyard.

And that is the point of pishing, Yankee stye. In the UK I think we tend to use it as a technique for bringing birds closer that we know — or at least suspect — are already there. We hear a tak or a tsip, and try to pish it out. In America, you pish and birds appear that you had no idea were there in the first place. At Cape May in November it worked brilliantly. One morning, I stood by an apparently empty hedgerow and pished shamelessly. Within five minutes I had a bushful that looked like the cover of a book on America's best loved birds: Cardinal, Chickadee, Tufted Titmouse, Carolina Wren, Mockingbird, Catbird, Brown Thrasher, Blue Jay, Ruby and Golden-crowned Kinglets, Bluebird, Downy Woodpecker and a Flicker. Not one by one, but all together. I kid you not. The bush nearly collapsed under the weight.

Ironically, the only thing that was missing was a warbler. But then, early November is well past peak warbler time in New Jersey. Except for Yellow-rumpeds, and there are still plenty of them. Mind you, at this time of the year, they don't have the dazzling black, white and lemon of spring plumage. The yellow rump is about the only bit of colour on them. Otherwise, they look like pipits. And they sound like pipits. And they constantly fly way up above your head like pipits. So how was I going to get a decent look at them? Easy: pish them down! Frankly, it seemed like a long shot, but it worked. Before my very eyes birds flying over, turned back, circled round me and finally plummeted into the same bush previously occupied by the great American selection, which had by this time realised I was just a demented British birder and pushed off (or should that be pished off?).

The rest of my Cape May stay, I pished wherever I went. It worked pretty well on American Sparrows: Song, Savannah, Field, White-throated, all popped out briefly. Whilst a quick pish in a coastal reedbed — whilst I was actually scrutinising dowitchers through my telescope — immediately produced four Sharp-tailed Sparrows, one of which literally perched on my tripod! In fact, just about the only birds that remained unpishable were the raptors. But then again, I'm not sure I'd fancy being enveloped by a flock of Turkey Vultures.

So why does David Sibley say that pishing doesn't work? Well, perhaps because he wants to find all the Cape May rarities himself (which he almost does, actually!) But mainly, of course, it's out of sympathy for the poor birds, who'd obviously be driven totally barmy if every one of the hundreds of birders who visit the Cape insisted on pishing them to distraction every few minutes. So, if you go to this wonderful place heed his plea: please keep pishing to a minimum. Mind you, I don't think the same restriction need apply in this country, if only because — in my experience — it really doesn't work very well here. Unless of course there happens to be an American bird in the woods. But how would you know? Tresco, Prawle Point, Porthgwarra, any Irish headland. Nothing about. Are you sure? Go on, give it a quick pish, just in case: it's always worth a try.

Dude birding
Arizona in April

I am always on the look out for holiday venues which will satisfy both myself and my utterly non-birding family. When they mooted the idea of a week on a dude ranch in Arizona I was optimistic. A new part of America — and a built-in birding pun: how could I say no? So I didn't.

Thus it was that, on the first morning, I staggered out of our casita (Mexican for extremely expensive swish villa, with 25" TV and air conditioning — dude indeed) feeling most delirious with jet lag, and wondering why I'd got up when it wasn't even light yet. It was 1 April. Mmm, and was I about to become the April fool? No siree. What followed was one of the best morning's birding I have every enjoyed. I shall attempt to convey it.

The first bird up was a quail. But not creeping around in crops as most quails do. This one was perched brazenly on the very top of a small dead tree, silhouetted against the rising sun. Gambel's Quail. And who was Mr Gambel? The man who designed the logo for the musical "Grease" perhaps? For that is what his Quail looks like. It's plumage is snazzy enough — very 1950s — but its crowning glory is an outrageous quiff, so enormous that any Teddy boy would shrivel in his drainpipes rather than attempt to compete. It looks like it's wearing a huge question mark. 'Impressed?' You betcha. Below the Quail, on the trunk of the same tiny tree, were two woodpeckers. The top one was a Ladderback — by name and nature — the lower one a Gila, which is probably another Mexican word that ought to mean some kind of cactus, since that's what Gila Woodpeckers love. The one I was watching proved it by flitting off the tree and landing on what looked like an oversized prickly cucumber, where it flattened itself against the spikes in a manner that brought tears to my eyes, but merely seemed to tickle the bird's fancy. It let out a merry laugh, as so many of the world's woodpeckers are wont to do. So did I. Three ticks already.

The sight of the cactus led me to ponder the nature of the surrounding landscape, which was by now bathed in a warm morning glow. Arizona desert. Not desert as in sand dunes and camels, but as in little green bushes, studded with bursts of wildflowers, and succulent with cacti of all shapes and sizes. Not desert as in deserted. Full of plant life, and...full of birds? You'd better believe it. Next one up was a big stripy thing with a not unfamiliar shape. Clearly a wren. On a cactus. It had to be Cactus Wren. The size of a Mistle Thrush, with the call of a Kookaburra. Wonderfully over the top. Another tick.

On I ambled, past the swimming pool and the outdoor jacuzzi (I told you it was dude) — two Hooded Orioles in the trees (vivid orange, black hoods, dazzling), and on down to the stables — real horses, an apparently real cowboy and, more to the point, a flock of mixed blackbirds, which of course, being American, weren't blackbirds in the thrushy sense, but glossy starlingy-type things: Red-winged (when they fly), Brewer's (shiny purple), Bronzed and Brown-headed Cowbirds (horsebirds in these parts) and Great-tailed Grackles. The latter begging an interesting semantic question. In Florida, I've seen Boat-tailed Grackle. This Arizona version is even more impressively endowed in the tail department. So 'great' is bigger than 'boat', apparently. So what's bigger than 'great'? Huge-tailed? Whale-tailed? This is just getting silly-tailed? Who cares? It was another tick.

After the stables, I played a hunch. About 200 yards away, I could see a sort of cliff. What would they call it out west? A bluff? A canyon, maybe. Not exactly grand, more a mini canyon. But a canyon nevertheless, with rocks. Surely the likely habitat of two more local wrens. Within five minutes, I'd clinched 'em both. On the rocks: Rock Wren, more normal wren size, and subtly brown and barred, as wrens should be. And in the canyon: you got it, a Canyon Wren. It was the song I recognised. Wispy and down the scale. 'That's a Willow Warbler', I muttered to myself. 'No it isn't', the bird replied. But it did sound like one, albeit amplified. A Willow Warbler singing through a megaphone: that's how to remember what Canyon Wren sounds like.

And, of course, that's what you have to do when you are birding abroad. If you don't actually know the local songs, at least you can relate them to British birds you do know. Like, for example, the European Wren song I was now hearing. Except it was coming from a warbler: Lucy's Warbler. They say that Americans aren't much into irony, but Lucy's Warbler certainly is. Most American warblers are delightfully colourful. Lucy's is grey. It really should be British. Maybe it was named by a comedian. Lucille Ball perhaps. I Love Lucy's Warbler...not. (Although it too was a tick.) But to compensate for the boring warbler, out buzzed a hummingbird. A male, sporting an imposing handlebar moustache, which glinted purple when the sun caught it. Costa's Hummingbird. Costa? The only Costa I can think of was Sam Costa, a British comic of yesteryear, who was famous for his...large handlebar moustache! (Honestly he was.) But surely they didn't name an Arizona Hummer after him? No, these were the ramblings of a jet-lagged birder, high on adrenaline and lifers. I moved on.

On down the path, every step brought new birds and new challenges. Sparrows: half a dozen different species; some dull — Brewer's, Song and Vesper; some dashing — Black Throated, Chipping and White-crowned. And even more eye-catching, but not a new bird for me, or for anyone who has visited America, a Cardinal. Or was it? The same extravagant shape, yes — crested and parrot-billed — but, on this one, the beak was ivory white, and the bird was grey, except that it looked as if its breast and face had been spray-painted with scarlet. Pyrrhuloxia! A bird or a disease? 'Did you get Pyrrhuloxia?' "Yep, but the ointment soon cleared it up'.

And so it went on. Three different flycatchers (king-sized kingbirds, actually), Curve-billed Thrashers, Black-tailed Gnatcatchers, Phainopeplas: daft names, splendid birds. And finally, just as I turned back towards the ranch for brekky, or 'chuck', or whatever cowboys call it, the bird I'd most hoped to see, but had strangely enough forgotten: Greater Roadrunner. And if you think the cartoon is funny, you should see the real thing. A ridiculous haircut, enormous feet, and the silliest walk this side of John Cleese. Sheer magic.

By the end of that first day, I had recorded a modest 60-odd species round the range, but no fewer than 23 of them were lifers. (And I've birded America quite a bit over the years.) By the end of the week, I had a not-a-lot trip list of 130, but 42 were new ones! All within an hour's drive of the Wickenburg Dude Range. Oh, and the family had a good time too. Yeehaaa!

Black Gold
The Increasing Price of Oilbirds

Earlier this year I was in Trinidad and Tobago. Just outside our Tobago hotel, on a big white wall, somebody had spray-painted — in red, green and gold, of course — the words: "It's nice to be nice." I'll drink to that, I thought, and indeed I did. Mainly rum punch, the national tipple in those parts. I also resolved that I would try and live up to that motto. Of course, I've failed frequently, but I'm going to have another go. Therefore this month I shall avoid being flip and facetious. Instead, I really will be nice. Mind you, I'm still going to start with a little niggle, 'cos its much more fun.

I was in T and T leading a small party on a leisurely-paced bird trip. Inevitably, one of the highlights of our week in Trinidad was a visit to the Asa Wright Centre. This is one of those almost legendary locations that most birders have heard of, even if they haven't been there. Those who have been there rave about it. They also rave about seeing the even more legendary Oilbirds. Oilbirds resemble large ugly nightjars — not that nightjars would win any beauty contests — and they are very rare. Oilbird is definitely a megatick; made all the more desirable by the fact that the birds try very hard not to be seen. They are exclusively nocturnal, and spend their days roosting deep inside pitch dark caves. The Asa Wright Centre is one of the few places that can arrange access to such a cave. At a price. You are not allowed to see the bird unless you spend at least three nights at the centre. The accommodation is not cheap. In birding parlance, Asa Wright is pretty 'dude'. The price of Oilbird is high.

I did hear the 'three nights' stipulation referred to as 'eco blackmail', but this is probably a bit strong. The reason for the rule was perfectly sound. A few years ago, the Oilbirds abandoned their cave. It was suspected that the problem might be 'visitor pressure'. So visits were abandoned too — which was logical, since there weren't any birds to see! However, as if to confirm the theory, the birds returned. Maybe, in the meantime, they had discussed the matter with the management and agreed to a less disruptive schedule. Anyway, it has worked. The birds are still there, but are now being visited by a smaller number of — 'paying' — customers.

That is, those who can afford to spend three nights at the centre. And therein lies the irony. No disrespect to ageing Americans with fistfuls of dollars, enormous lenses and incredibly loud voices, but I do suspect that some of them wouldn't know the difference between an Oilbird and Donald Duck. But they've got it on their lists.

Meanwhile many genuine, less affluent — do I mean British? — birders haven't. Is this right? (Or do I mean Wright?) After all, who do Trinidadians relate to better? Us or them? Does America play the West Indies at cricket? And Dwight Yorke turns out for Manchester United, not the Harlem Globetrotters. Or am I sounding bitter?

Maybe, but actually I can live without Oilbird — well, I've got to, haven't I? But I am concerned that this kind of thing could catch on over here. Imagine if the next time a Parula Warbler turns up on Tresco, you're not allowed to land unless you buy a week's time-share in one of the estate cottages. Or Tesco, even more so. "Waxwings in the car park? You can only look at them if you purchase £50 worth of groceries." Oh no, I'm sorry, I'm being flip and facetious.

Get back to being nice. And get back to Trinidad. Look for consolations. Well, presumably those Oilbird dollars are paying for the upkeep of the Centre and the conservation work that goes on there. So that's good, What's more, it is possible to pop in to Asa Wright for a day, for a very reasonable entrance fee, and join the gentry on the balcony as they sip their freshly ground coffee and tick off a quite incredible array of hummingbirds, tanagers and so on that are lured in by a veritable phalanx of constantly replenished artificial feeders. I wouldn't call it exactly challenging birding and I'm not sure it is entirely 'natural' — do Chestnut Woodpeckers usually eat sliced mango? — but it is a lot of fun. And, be honest, it's nice to be a dude for a day.

However, if you fancy birding Trinidad — and I do recommend it — on a slightly less luxurious level — and here's the really nice bit — may I recommend the Pax Guest House? It's up in the hills, conveniently near to the airport and lots of the best birding sites. There's a great view, excellent birds, lively food and it is run by a gentleman who is not only a decent birder, but also — better still — understands and caters for birders' needs and habits, including early morning starts, packed lunches and reliable guides who, by the way — for a little extra — might just be able to take you somewhere else where you can tick off a certain cave-dwelling ugly bird.

No, I'm not being paid to say this. It's just that we had a really lovely time at Pax and it deserves to do well. That's why I'm saying it…Mmm. You know, they're right, it is nice to be nice.

BIRDS AND WORK

Well, birds and my work, I suppose I should say. I was fortunate enough to be filming in Trinidad and Tobago for the second series of Birding with Bill Oddie (BBC, TV). Working on that series also inspired several other pieces. It also provoked one rather grumpy letter, defending the Welney meanie (see 'Value for Money', page 47) and accusing me of rampant hypocrisy since, in effect, the BBC paid me to go and chase rare birds. Well, I guess you could put it that way!'

Rare birds going cheap
On getting ticks for free

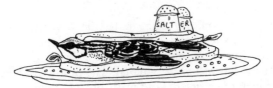

Judging from the flood of letter (one) that I have received, it has become apparent that there is a certain resentment out there that the BBC actually pay me to go twitching. I do feel the criticism a bit unfair. After all, any programme about birdwatching is quite likely to involve looking at birds, and surely I can't be expected to close my eyes if a rarity happens to turn up. What's more, anyone who has seen the series will appreciate that in fact rarities are largely (co)incidental to the main themes. (And more's the pity, some of you may well think!) I admit we did wilfully and intentionally film four Cranes at Minsmere two years ago, the Blue-cheeked Bee-eater in Shetland last June, and a Pallas's Warbler at Cley in October, but the two other official rarities we happened to come across both ended up on the cutting room floor: Black Kite (too tatty) and Canvasback (too boring). So I really don't think I should be accused of being paid to chase rarities. Mind you, I suppose

I should confess that I do have an arrangement with the beeb that they will transport me by chauffeur-driven limo to tick off anything, anywhere, any time for my own British list, and I do have a personal masseuse (doubling as occasional concubine) to relieve the stress of dipping out. But that's your licence money well spent I'd say, and I'm sure you wouldn't be so churlish as to begrudge me my little perks.

But seriously though, presenting wildlife films involves far more 'ptcs' (pieces to camera) and 'travelling shots' (I recently spent four hours trundling round the M25 to provide a 10-second 'link') than birdwatching. I do try and make a bit of a fetish of talking about the birds as I'm actually looking at them (which is more than can be said for some natural history progs I could mention, but won't, for fear of offending any other presenters), but it's the cameramen who get to see most and get the best views. Anyone who has done a bit of in-front-of-the-camera stuff will surely confirm that it's mostly really really boring. Don't get me wrong, I truly love my work, but, ironically, I was able to do more 'real birding' during filming back when I was doing comedy shows rather than in more recent years.

Which brings on a flashback to the Seventies (OK now apparently, 'cos it's 'retro' or 'cult' or something trendy). This was when I spent large chunks of the year shooting The Goodies (younger readers ask your parents). The producer decreed that it would be more efficient (and probably more fun) if the whole unit went and stayed in a hotel away from London for maybe two or three weeks at a time. The criteria for where we went were two-fold: firstly, that it provided a variety of locations as required by the scripts, and secondly, that it was a good area for birds. I'm not sure the BBC realised the second factor, but since I co-wrote the scripts I felt they need never know. Anyway, the outcome was the various series were filmed in Bognor (near Pagham Harbour), Folkestone (near Dungeness) and — most memorably for me — in southwest Cornwall and on Portland Bill. During the Cornish fortnight I not only had a Richard's Pipit flying over during a tea break, and a glimpse of a mystery warbler late one evening at Porthgwarra (which to this day I suspect was an Orphean), but I even managed a day trip to the Scillies, during a lull in the filming schedule. This was of course a very irresponsible thing to do, as I could easily have got stranded there and cost the BBC thousands of pounds. In fact, it nearly happened.

I was hurrying back to the heliport to catch the last flight (only mid-afternoon in those days) when I heard tell of a Pallas's Warbler in Holy Vale. It would have been a new bird for me so I made the instant rash decision to go for it and catch the Scillonian instead. I ran to Holy Vale, where two or three birders were contentedly peering at a large sycamore, presumably enjoying the Pallas's. I raised my bins and followed the line of their gaze, just in time to see the merest flicker of a lemon rump dive into the canopy and disappear as rapidly as a stone from a catapult. 'Don't worry', I was assured, 'it does a little circuit. It'll be back soon'. But an hour later it hadn't returned, and I had barely 20 minutes to race back to Hugh Town to catch the boat. I may well have set two birders all-comers athletics records that day: one for the two-mile race to the harbour, and the other for the long jump I had to make leaping from the quayside onto the Scillonian, five seconds after they'd pulled up the gangplank. It wasn't a relaxing experience, made worse by the fact that I surely couldn't count half a Pallas's Warbler (especially not the rear end).

As it happens, I got the other half some years later when we were filming on Portland. It had been an excellent week (birdwise, I mean; and the comedy wasn't bad either). The terror of being bombarded by an avalanche of exploding eggs ('Goodies and the Beanstalk') was considerably alleviated by a superb movement of winter thrushes alongside our giant pantomime geese; and an Alpine Swift flew over during a Buster Keaton parody. Good job it was a silent movie. Lip readers have studied the clip and deciphered my expletive as 'f**k me, it's an Alpine Swift!'. (Not one of Buster's best known catch phrases.) Then, one evening, I nipped down to the Portland obs and ticked off a nearby Yellow-browed Warbler. There were obviously good birds around, so I set up an arrangement by which the obs would ring our unit manager and he would pass on any news to me as soon as I had a break from filming. Fine in theory, but a recipe for 'Chinese whispers' if ever there was one. I should have known. It had happened to me before. I remember my wife passing on a call she'd taken telling me there was a 'Cheese Sandwich Warbler in Scotland'. Chestnut-sided, would you believe. Thus I should have been wary when I was told during the lunch break: 'Oh the bird people rang. They've caught a Crested Warbler.' 'No such thing', I muttered. 'Are you sure it wasn't a Yellow-browed?' 'Oh yes, that'd be it.' 'Mmm. Nice. But I don't need it.' So I didn't go.

OK, I agree, 'Crested' doesn't sound much like 'Yellow-browed'. But it doesn't sound much like 'Pallas's' either. When I finally sauntered down to the obs that evening, the warden was almost disgusted at my nonchalance. 'Why didn't you come at lunchtime? We kept it as long as we could.' I had a choice between tears and panic. 'Where is it now?' 'We let it go in the garden.' I raced outside and, in fading light, that's where I glimpsed the other — stripier — end of a Pallas's. It was another 10 years before I saw a whole bird, and another 10 before I got a really really good long close-up view. That was last October at Cley. It was great. And, what's more, the BBC paid me for it. Ooops! Now I suppose I'll get another letter.

Capper caper
How can you miss it?

I shall always be grateful to Capercaillies. If only for inspiring the most memorable headline to a learned ornithological treatise...ever. A few years ago, *British Birds* attempted to up its circulation by brazenly emblazoning the question across its front page (or was it page three?) 'Why do capercaillies have such big cocks?'. I don't know if this blatant attempt to lure a gutter readership worked, but for weeks a single copy of BB lurked on the top shelf at my local newsagent, peeking out from between Forum and Health and Efficiency. Mind you, Capercaillie cocks may well be big, but not big enough I'd say. The fact is, I've never seen one. You see — despite the conclusion reached by the BB article — size isn't everything. The Lock Ness Monster is reputed to be ginormous, but I haven't seen that either.

However, I do have Capper on my British list, but it is a fading memory. Long long ago, early 1960s I think, I did the Spey Valley circuit to tick off the Scottish specialities. I recall one day, deep in the Rothiemurchus Forest, I'd been caught short and had nipped behind an ancient pine to relieve myself, when I was suddenly scared out of my wits by a female Capper clattering away from almost under my feet. I nearly evacuated more than I'd intended. She was followed by what appeared to be half a dozen panicking Partridges, but they were, of course, baby Cappers; probably barely out of the egg but already capable of flapping through my legs at a dangerous height, both to my dismay and pleasure. But, though I wandered the forest for much of the rest of the day, I never did catch up with the bid daddy.

It was 30 odd years before I tried again. In June 1996, I was filming for my forthcoming TV series (small plug coming up) Birding with Bill Oddie (imaginative title, huh?) which may be on your screens at this very moment; or then again it may not (the BBC have a cavalier programming policy matched only by the publishers of bird books). The idea was to do a Scottish weekend and capture all the target species on video. Did we succeed? Why not watch and find out (as the BBC publicist would like me to say). But I'll tell you this much: we didn't get one species.

Ironically, we'd thought it was going to be easy. For the past couple of years, there had been a rogue male Capper terrorising visitors to the RSPB reserve at Abernathy. Apparently, it was in the habit of actually attacking twitchers (a bird of taste?). It was incredibly easy to photograph, as long as you clicked the shutter before it swallowed your camera. It had even been reported leaping onto vehicles and pulling off windscreen wipers, like the baboons at a safari park. Clearly the RSPB had got it very well trained indeed. Or had they? Well, the irritating fact was that, just as we were about to immortalise the delinquent on the telly, it disappeared. No one seemed to know exactly what had happened to it. Had it died? Had it been deported? Or maybe it was in prison awaiting trial for assault. Or had it got a more lucrative job as a guard bird with Securicor, or as a bouncer at the Glasgow Empire? In any event, it was now, to all intents and purposes, an ex Capper.

So, we would have to find our own. Filming schedules are tight and time is short, so clearly I needed help. This was provided in the form of a guide. He shall remain nameless, but I will tell you that he was Scottish. Now, far be it from me to stoop to facile racial stereotyping, but it does seem to me that Scots do often come in one of two types. There are jolly, optimistic ones, like Billy Connolly and Andy Stewart, and there are rather dour, philosophical ones, like Ivor Cutler and...well, like the chap who was supposed to show me the Capper. I knew I wasn't in for a cheery dose of the Billy Connollys from our first exchange. Me: 'Oh this is brilliant, I'm going to see a Capper.' Him: 'You'll be lucky.'

Well, I wasn't. To cut a long story — and quite a long film sequence — short, we drove round the forest tracks in a Land Rover from 3.30 in the morning until 10 o'clock, during which time I can honestly say that I never once felt that there was the slightest chance of us seeing a Capper cock, no matter how big they were. It wasn't that they weren't out there somewhere. My guide assured us that they were, but he equally assured me that they were not numerous, extremely wary, not displaying at this time of the year and so on. He was obviously right, and to this extent he demonstrated how knowledgeable he was about the birds. His pessimism had an air of totally convincing authority.

But why did I begin to develop this teenie niggly feeling that he was almost enjoying my growing frustration? Maybe it was the way he teased me by every now and then screeching to a halt as if he'd spotted something, only to leap out of the Land Rover, scoop up a handful of dropping and announce: 'Well, this proves they're here somewhere.' Which, in at least one instance, it didn't, since he then sniffed the droppings and muttered: 'Dog!' On another occasion, he showed me what looked like an ancient tyre track, and assured me that it was in fact a Cappers dust bath. Which was surely true, since there was a single manky minuscule feather in it. (It could have come from a Dodo as far as I could tell, but I took his word for it.) As the search continued, to make me feel better — or worse — he consoled me with gems of Highland wisdom.

Me: 'I'd love to see a male Capper'. Him: 'Well Bill, you can't have everything you want'. Me: 'Wouldn't it be great if one hopped out now.' Him: 'Life's not like that Bill.' And, just to finish me off once and for all... Me: 'Well, its been over six hours. I guess we give up. Thanks for trying.' Him: 'Well actually, Bill, I'm quite glad we didn't see one. That means they're safe on their nests.' Me: '*****!' (Sorry, mustn't use naughty words. Leave that to *British Birds*.)

Seriously though...it was fun trying. And, as indeed my guide pointed out, I have an excuse to go back to a wonderful part of the country. And of course the birds are more important than telly programmes. And did we leave the Capper hunt in the final cut? Well, why not watch and find out.

Telly twitching
Shetland delivers

I don't get a lot of British ticks these days. Not, I hasten to add, because my list is so long (in case you're curious, its probably no more than about 370, but I don't honestly know). It's just that my general philosophy on twitching is that I won't travel far, but if I do 'happen to be in the vicinity' then I'll go for it. Imagine my surprise, delight and luck then, when I recently had two new birds in two days. In late June/early July I was up in Shetland shooting an episode for Birding with Bill Oddie series two (due out next spring). As we picked up our car at Sumburgh Airport it became evident that neither I nor my producer (Stephen Moss) were entirely unaware of the fact that there was a Blue-cheeked Bee-eater within half an hour's drive. It wasn't actually in the script, but we reckoned we might as well go and film it. So we did. So, specially for those unfortunate twitchers who flew all the way up to Shetland, dipped, and flew back a few minutes before the bird did…tune in next year to see stunning pictures of Blue-cheeked Bee-eater so close that you can see the panic on the bee's face. Well, that's one way to get a few more viewers!

The next morning I got my second tick. This time driving across Yell on our way to catch the ferry to Fetlar. There had been a Black Kite lurking around for some time. We just happened to be passing the place it had been most recently seen. There were a couple of birders scanning, so we slowed down, and immediately spotted the bird, flapping around conspicuously, as if it was in its contract to provide a target for every species that fancied a bit of mobbing. One by one they took it in turns: Hoodie, Lapwing, Oystercatcher, Raven, Redshank. Fun for them, and an instant twitch for me. Just the way I like 'em. I confess Black Kite isn't really my favourite species, I've seen millions of them abroad, and they always look scruffy and are usually rummaging around over — or even on — some nasty, smelly, rubbish tip. I've missed several in Britain —

'you should have been here five minutes ago' — but it didn't really bother me. 'I wouldn't cross the road to see one', I said. Well, I didn't have to! I just leaned on the car.

The kite we didn't film, but that was OK because, after all, we weren't in the Northern Isles in late June to chase rarities. We were there to capture the 'magic of Shetland'. Thus I was duly dive-bombed by Bonxies, dazzled by Red-necked Phalaropes, and charmed by the most approachable Puffins outside of a glass case at the Natural History Museum. The magic of Shetland indeed; and that's what most of the programme will be about. I hope you will watch, but I'm not going to go on about it here. After all, be honest, the reason I really enjoy writing my *Birdwatch* piece is that I can be a bit more esoteric, self-indulgent perhaps, even a bit twitchy in my own way.

Which takes me back to the rarities. The fact is that, to me — and I dare say to you, dear reader — a very large part of the magic of Shetland is that, even in late June/early July, the most extraordinary birds do turn up there. Some of them the locals almost take for granted, like the male King Eider we nipped down the road to admire, whilst waiting for the Blue-cheeked to return to its garden. And yet, call me perverse if you like, the birds that really got me excited weren't the pukka rarities — all of which lacked that final frisson because we knew they were there — no, it was the oddities (pardon the pun) that we just happened to come across. For example, a pre-breakfast vigil in a Lerwick garden, when a male Siskin suddenly appeared on top of a sycamore and sang its heart out, until my eye was distracted by a movement under the canopy which turned out to be a Hawfinch, which then treated me to arguably the best view I've ever had of one. Hawfinches aren't common on Shetland let alone in a garden sycamore. It was the incongruity I like so much. And talking of incongruity, I loved the moment when we were filming Gannets on the Noop of Noss — edge of a sea cliff, not a tree within miles — and what suddenly flits across behind us? A flock of 13 Crossbills! Even the Bonxie stopped harassing me and stared.

Of course, the Crossbills fitted into a much wider picture. And that's another thing I love about birds: the times when patterns emerge and connections start making sense. (Except that the whole thing is wondrously mysterious anyway.) Like this summer's Crossbill invasion. With perfect symmetry, the day before we flew off to Shetland, I had had a flock of 13 Crossbills flying over Hampstead Heath. (Obviously the same ones I saw on Noss. I recognised them.) That day, I'd called Birdline South East and listened to a veritable list of other Crossbill sightings across Kent and Hertfordshire. How satisfying: local patch fits into regional context, which fits into national pattern.

As I write (mid-July) it's still going on: yesterday, I had a flock of 30 Crossbills over the Heath. Which set me off on Crossbill memories! Like seeing my first ones at about a mile range, chipping over fir trees near Sidmouth. I had to take my companions word for it. 'They are Crossbills, honestly.' And how many times have I seen them really well? All of my sightings this summer have been in flight, noisy and conspicuous, and yet slightly frustrating. It's like a birding mate of mine says: 'Crossbills: they are either there, or they're not'. I know what he means. Which reminds me of one lovely bird that was very much there. October, 1990, Flamborough Head. East wind blowing, Redwings and Goldcrests cascading along Old Fall hedge. I spotted a scarlet fluff-ball nestling by the side of a path. I confess my first outrageous thought was: 'Tanager'! As we crept closer, it revealed itself to be a male Crossbill. It didn't move. Dead? No, fast asleep. Exhausted, after arriving from...well, where? We were able to catch it by hand. Its measurements revealed it to be too big for a Common, yet too small for a Parrot. Theoretically, it fitted Scottish. But I think not. Presumably there is a size cline across the European range; and yes, I'm very sceptical that there are three separate species. But, frankly, I don't really care. I'm just glad they're here.

And if this month' piece seems like a rambling stream of consciousness and connections...well, isn't that exactly what is so enjoyable and endlessly fascinating about birding?

Friend or foe?

What a wondrous piece of work is man

You know those natural history programmes about 'predators', that feature more blood and biting than the front row of a rugby scrum, but always end up with a doomy commentator saying 'but the greatest predator of all is man', and then slag us off — yes, you and me, man and woman — 'cos we're ruining the environment and killing all the birds? Well, I'm getting really fed up with being attacked like this. So, on behalf of my species, I hereby wish to defend humanity and ask: 'Where would the birds be without man?'. Floundering around with nowhere to perch, feed or nest, I'd say.

I came to this heretical conclusion last month when I was filming for Birding with Bill Oddie in Israel. We started in Eilat, a Red Sea resort whose ambience and architecture would make Port Talbot look quaint (which believe me it isn't). And yet this wondrously ghastly and totally unnatural oasis is a veritable magnet for migrants. In spring, there are warblers on the washing lines, flycatchers in the flowerbeds, and buntings by the beach huts. Even in mid-January we found a Red-backed Shrike hunting from a kid's climbing frame, and a Rufous Bush Robin pulling worms out of a lawn softened by a garden hose. The birds come to Eilat not in spite of what man has done, but because of it.

The Lichtenstein's Sandgrouse site was, and is, an even better example. Eilat regulars will know it well. There is a water pumping station up on a hill just behind the town. It looks like a cross between a small modern sewage farm and a prison: barbed wire, concrete buildings, a horrible noisy pump, and a lot of pipes. One of them used to leak, and the drip soon became the traditional evening drinking fountain for the sandgrouse. The pipe doesn't leak any more, but the enlightened local council have syphoned off a tiny little permanent pond, and the grouse still slake their thirst there every evening. One might wonder why. It isn't as if there aren't plenty of other sources of moisture near the Red Sea, but there must be something about the Eilat municipal water that is the very elixir of life to Lichtenstein's Sandgrouse. They are clearly grateful. So are the birders.

Our next location was a little more out in the wilderness. A 'wadi': a parched river bed snaking its way into a rather gorgeous rocky ravine, fringed with acacia trees. Very pretty, very natural. And where were the birds? In the car park! A Desert Lark shuffled around in the shade provided by a Peugeot 305. A Blackstart nibbled on the remnant contents of a discarded crisp packet; and a White-crowned Black Wheatear investigated the crumpled front page of an old Hebrew Newspaper (I dread to think what that had been used for).

And so it went on. Further north, we filmed Great Black-headed Gulls at a fish farm, and Bluethroats by a cesspit. Even the raptors backed up my theory. There was a fantastic selection, but where did we see most of them? Swirling majestically over the natural peaks and valleys of the Negev? Nope. They were perched like giant budgies along a row of whacking great electricity pylons. One scan revealed two Saker Falcons, three Long-legged Buzzards and an Imperial Eagle. Not to mention the Kestrels, Peregrine, Merlin, Black Kites and Hen and Pallid Harriers mooching around the same area. And why were they all there? Because of the easy food supply. Top of the menu: flocks of hundreds, or even thousands, of Skylarks. And why were the Skylarks there? Because of the fields were furrowed and planted with lots of lark food. And who provided that? The greatest predator of all, apparently.

But then birders have known this for years, haven't we? Let's face it, you only have to consider the favourite habitats of just about any bird you can think of to realise they are man-made. Rubbish tips for gulls; sewage farms for waders; telegraph wires for shires; fences and walls for chats and flycatchers; vicars' lawns for Hoopoes; supermarket car parks for Waxwings. And what about the nest sites we provide? What on earth did our best loved garden birds do before we put up lamp posts and letter boxes for the Great Tits, and started chucking away old kettles for the Robins? Not to mention the species that are so dependent on us that their names admit it: House Martin, House Crow, House Finch, Barn Swallow. It's about time we changed a few others, if you ask me. How about Gravel Pit Plover (instead of Little Ringed)? Motorway Hawk (for Kestrel), Millennium Dome Chat (for Black Redstart).

'The greatest predator?'. Maybe. But think about this: OK, so humans poison, pollute, hunt, and destroy habitats, but it seems obvious to me that if it wasn't for man there probably wouldn't be any birds in the first place! You see, we giveth, and we taketh away.

Oh, before I start getting letters...only joking. Well, sort of.

Over the rainbow
Beginning to see the light

Everybody knows that when you take a photo you should get the light behind you. Shoot into the sun and at worse you'll get a complete silhouette, at best a psychedelic effect of 'donuts' and prisms. What you won't get is true colours. Everyone knows it. But I sometimes wonder if birders apply the same logic when they are looking through binoculars. I've often seen people squinting into the sun and struggling with identification. It seems particularly daft at a reserve surrounded by hides, or at a reservoir with a road round it, or in a wood. At the risk of sounding appallingly patronising, I feel like offering a simple piece of advice: try moving! Get the sun behind you. Or, if you can't do that, at least shift a pace or two so the bird is against a dark background, not a glaring sky. A Black Wheatear against a horizon becomes a blue and peach Northern against a rock. A Black Woodpecker becomes a Green against a tree trunk, and a 'possible Ross's Gull' is revealed as 'only a Black-headed' when it dips down against a cliff face. Anticlimactic conclusions I grant you, but at least you can pride yourself on your field craft. But I'm sure you do already. Sorry.

So what got me thinking about this lighting business? Well, we were in Holland recently (Birding with Bill Oddie, again) when, ironically, Andrew, our wildlife cameraman, suddenly started doing exactly what I've just said he wouldn't: he started shooting straight into the sun. I immediately began commentating (this is how we do a lot of the programme, 'as live'). 'A birder would never look at birds this way', I said. Then I asked Andrew for a quick play back of the tape and added: 'but perhaps we should'. The image was exquisite: a flock of Snow Buntings fluttering along the shingle. Silhouettes, yes, but with magically translucent wings. 'They're like little angels', muttered Andrew. I added that to the commentary too. And it made me think: How often do we birders switch off the obsessional ID button and tune in to the purely aesthetic? A pretty picture, a memorable image, a stunning lighting effect. We've all seen them. But do we actually look for them? Andrew does, and I guess I'm particularly fortunate to often have him with me to capture them on video tape.

For example, a couple of such moments from our shoot in Israel this January. Gulls against the sunset. For a kick off, this had the extra plus that it obliterated the disquieting but undeniable fact that the vast majority of eastern Yellow-legged Gulls — yes, cachinnans, the one they call Caspian Gull — didn't have yellow legs at all. In good light they varied from yellowish to pink, via various shades of grey. Against the light, however, they were a deep golden bronze, as indeed was the rest of the bird. The effect was splendid, especially when one individual positioned itself precisely in front of the blood red orb and was thus surrounded by a fiery halo so majestic that Andrew instantly christened it 'The gull of God!' Eat your heart out Jonathan Livingstone.

But if we thought that was a memorable image, the next evening surpassed it. We were in the Hula Valley, hoping to film thousands of Cranes coming in to roost. Alas, it had been bucketing down all day. The birds would still come, and we could still take pictures, but they would be of grey shapes against a grey sky. Still impressive, but definitely not at their best. We were disappointed, but our schedule simply wouldn't allow us to try again the next evening. However, as we splashed down the muddy track to get into position a veritable vision unfolded. Suddenly, the lowering cloud cover parted in front of our Ford Transit, as miraculously as the Red Sea had done before Moses (sorry, but you start thinking like this in the Holy Land). And lo! A mighty shaft of light illuminated the valley. The background was still inky black, but the foreground was bathed in sunshine. Then, as instantly as if the great lighting man in the sky had switched on a couple of laser beams, a double rainbow appeared. The Hula Valley had become a gleaming amphitheatre framed by a multicolored proscenium. The stage was set. Enter the players. Several thousand Cranes took to the air, flapped through the spotlight in natural slow motion, glided underneath the arches, and cascaded down, literally at the end of the rainbow! You expected Jean Michel Jarre music to blast out, except that it would have been drowned by the far more magnificent bugling of the birds. The whole event lasted barely two minutes, before the cloudy curtains closed again, and a great darkness descended onto the land once more. And no, I didn't have my camera with me. But Andrew did. But did he manage to tumble out of the van, set up his tripod, and focus his monster lens in time to capture that once in a lifetime image? Watch your TV screen sometime early in the new millennium. Beats that stupid dome any day, I'd say.

On the slide

Don't trust the Trust!

This year I'm doing quite a lot of live shows, so, I recently took delivery of a smart new Leica slide projector. I shall always carry it in the boot of my car, 'just in case'. In case of what? Well, this sort of experience.

November 1998. I was due to do a show for a local wildlife trust (who are lovely people doing a great job, but probably prefer to remain anonymous!) I arrived in the foyer of the hall at about 6 o'clock to be greeted promptly and efficiently by the organiser. 'I've arranged for some sandwiches in the pub over the road.' So far so good. (It's amazing how often organisers don't consider the fact that if I've driven for several hours and have to do a show that won't finish till 10 o'clock, after which I then have to drive back for several more hours, I just might fancy a quick nibble at some point.) On the other hand, first things first. 'Great', I replied, 'but I'd just like to check the technical stuff.' 'Oh, OK.' Did I detect a twinge of nervousness as he led me into the hall? Well, I say hall, gymnasium more like. There were ropes, wall-bars, exercise mats. 'So does the audience sit on the floor, or dangle?' 'It's OK, we put out chairs.' And indeed, even as he spoke, a sort of enormous prefabricated grandstand was being trundled into place. 'Fine. Where's the screen?' Heaven knows, I've managed to project onto just about everything from lace curtains to a pocket handkerchief, but a climbing frame? I thought not. But no worries. A turn of a handle, a spin of a pulley, and down from the ceiling unfurled a screen vast enough to have done justice to Ben Hur. 'OK then, sandwiches?' urged my host. 'Nearly. Just one more thing. Can I try the projector?' 'Ah,' sighed the organiser.

Oh, how much doom can be conveyed by one theoretically meaningless syllable! 'Ah.' You just know it's the preface to a pronouncement that is about to shatter your life. 'Ah.' It could be uttered by a doctor, or an bank manager, or a partner, but nothing can strike dread into a guest speaker's heart like the 'Ah.' of the organiser of a slide show. Especially when it is followed by the words: 'There's a bit of a problem with the projector.' 'Oh yes?' I winced. 'What problem's that?' 'Er, we haven't got one! But we're working on it. Meanwhile, shall we go and eat?'

It's amazing how panic obliterates hunger. Eat a sandwich then? I could have as soon nipped to the touchline for a snack before taking a penalty in an FA Cup Final. 'Haven't got a projector!' I spluttered. The organiser spluttered back: 'We thought the hall had one, but they thought we were bringing one. But they haven't, and we haven't.' 'But I can't do it without a projector', I protested, stating the obvious, but supporting it with an analogy: 'It's like booking a recital and telling the pianist 'Sorry we haven't got a piano. Maybe you can hum a couple of concertos!' 'I know, but it's OK, we're getting one', assured the organiser. 'So shall we go to the pub?' 'No. It's not OK', I snapped 'I'm not leaving this hall, gym, whatever it is, till I see a projector. And if I don't see a projector, I am leaving, forever.'

We waited for about half an hour. Then, in staggered a bloke carrying a large cardboard box, which I swear he was blowing dust off. Out of it he produced a beige plastic contraption so ancient that it would have given Arthur Negus a hot flush on the Antiques Road Show. It was called a Goblin something. 'That's a Teasmaid, isn't it?' 'No, it's a projector,' said the bloke, ignoring my sarcasm, but adding apologetically: 'It's been in my attic for about 20 odd years.' 'So, does it work?' I enquired. 'We'd better try it.' So we did. And it didn't. My glass-mounted slides wouldn't fit in the slots at all, whilst the cardboard ones either jammed or shot out all over the gym, like tennis balls from one of those practise machines. 'That's it.' I announced, invoking my analogy again. 'This is like telling the concert pianist 'Sorry, but can you play Rachmaninov on a Casio with half the keys missing? The show's off.' 'Fair enough,' agreed the bloke 'but what about the 300 people waiting outside?' Eh? I looked at my watch. It was now 7.30. Time to let in the audience for an 8 o'clock start. And 300 people had come to see me. I'm a hard man, but vanity alone made me loath to reject a crowd that large. At that moment, the organiser saved the day. Or so it seemed. He rushed in brandishing his mobile as if it were a magic wand: 'I've managed to get hold of the director of the Trust. He was just leaving his office to come here. He's bringing the Trust's own projector. State of the art. He'll be here by 8 o'clock.' 'Fine,' I said 'Now I'll have that sandwich.

Five to eight. Back we came from the pub to find 300 people seated, and a few more hanging from the wall-bars. But no director, and no projector. 'Traffic must be bad. He'll be here.' 'OK, we'll wait a bit.' We did. Then we waited a bit more. We waited quite a lot. Nearly 8.30. I hate it when 'performers' keep audiences waiting, so I went out and explained what was going on. Or, rather, what wasn't. I got a laugh by suggesting that the audience could pass each slide round and look at it with a torch. Then I brought forward the raffle from the interval, and dragged it out by making each winner do an acceptance speech. I was just about to pick teams for charades when in galloped the director. Flushed, harassed, and muttering about being stopped by the police for speeding, which I was, of course, ungracious enough to tell the audience, and get another laugh. I also told them the good news; We now had a projector. And the bad news: it was the 'wrong' sort.

My slides were in carousels, this one used straight trays. I offered the audience a choice. They could take a 10-minute bar break, or sit and watch me switch nearly 200 slides into new trays. They unanimously agreed that they'd like to watch, presumably in case something else went amusingly wrong. How right they were. Whilst I fumbled with slides, several other people tried to set up the new projector on a little table in front of the huge screen. Unfortunately, straight-tray projectors can't be tilted as much as carousel types, otherwise you can't get the tray in the back (think about it, honest, it's true). Thus began the great tower-building routine. Urged on by suggestions from the audience, the organiser, the Goblin bloke, and the director stacked a chair on top of the table, two telephone directories on the chair, the Goblin box on the directories, and what I think was the top section of a vaulting horse at the summit. On top of that was balanced the projector, swaying precariously, but at least beaming a bright rectangle onto the screen high enough for all the audience to see. All I needed to do was put in the slide tray. Alas, the projector was now about four feet above my reach. This was clearly a job for the director who was as tall as a basketball shooter, though even he had to stand on a chair. But at last I was able to start — just. To add one final farcical touch, the slide changer was not a remote control, but a very short cable. It dangled down just far enough for me to reach it.

As long as I stood right by the 'tower', on tiptoe. I certainly couldn't move around, and if I risked turning to look back at the screen I was likely to yank the projector down from a great height, probably wiping out several people in the front row. But I managed, and the audience were lovely, and fun was had by all — eventually.

Nevertheless, if you ever come to one of my shows, I promise it will never be like that again. 'Oh, spoilsport!' did I hear you say?

BIRD RACING

As I'm sure you know, this doesn't refer to homing pigeons or duck derbies. Bird races involve teams of birders (usually in fours) tazzing around trying to record as many birds as possible in 24 hours, usually sponsored, and thus raising stress levels and a lot of money for conservation. For a full account of such an event try and track down a copy of The Big Bird Race (Oddie and Tomlinson. Collins. Out of print, but they turn up in old book shops and the like!).

Meanwhile, here are a couple of pieces from *Birdwatch*, and a longer account written for the Hong Kong Bird Club.

Confession time
Fair play on bird race days...or is it?

This May, I joined the official *Birdwatch* bird race team for its annual 24-hour dash round the London area. We didn't do terribly well this year, but I can claim that we were scrupulously honest.

This is the thing I'm always asked whenever I explain bird-racing to non-birders: "How do you know that people don't cheat?" Naturally, I reply that birders are far too honourable to do such a thing. And even if we aren't, there's a sort of collective conscience (or embarrassment) that prevents us from transgressing — throughout the scores of bird races I've taken part in, I have never been aware of any wilful cheating.

On the other hand, in the course of a typical bird race there are likely to be several, er, how can I put this, 'moral dilemmas'

to be resolved. Heaven knows, I'm not confessing anything —
and I'm certainly not accusing anybody — but...well, let me just
remind you of a few such situations. And if your ears start
burning, don't worry, mine are quite hot too (it takes one to
know one!).

Noises in the night. It's 2 am. Pitch dark. You're standing at
the edge of a wood next to heathland. You're listening for Long-
eared Owl. You try calling them up. A nice clear hoot answers.
"LEO! We'll have that!" OK, you saw the headlights of another
bird race team heading for the other side of the wood, and it
may well have been their owl impressions that you've ticked off.
But then they've probably ticked off yours, so fair enough?!

And what about those other nocturnal aural temptations?
Barn Owl. Irresistible. Just look up its calls in the book:
"shrieking, hissing, snoring". All noises readily emanated by
other denizens of the night: climaxing courting couples, vandals
letting down tyres, even dozy members of your own team. It's so
hard to be sure of the truth in the dark, isn't it? Was that a
booming Bittern, or a pregnant cow? A migrating godwit, or a
goat? And yes, we all know young Long-eared Owls and
Nightjars sound like squeaky gates and motor scooters...but so
do squeaky gates and motor scooters!

Mechanical noises can still sound promising by day, but it's
harder to kid yourself. I've been present when a Golden Oriole's
calls were reidentified as a boy with a bicycle pump; and when
a reeling 'Grasshopper Warbler' was finally traced to an electric
pylon! We owned up...but does everybody?

And talking of 'Groppers'...these often feature in the 'I'm
sorry, I can't hear them' scenario. So what do you do when one
of your team members confesses that he is deaf to certain
frequencies? I recall having to carry one such limited listener to
within a yard of a reeling Gropper, and finally drop him on the
bird. He claims he did hear it yell! Other people can't hear
Treecreepers, or tell the difference between the various sips of
tits and 'crests. In such cases, the temptation is to invoke the
'you must have heard it' line of reasoning. This goes: "The bird
called, we three heard it, you were with us, so you must have
heard it too." (It's just that you weren't aware of it!)

The same applies to the 'you must have seen it' syndrome. An
example: your team is scanning a flock of Dunlin on an estuary.
Someone gets onto a single Curlew Sandpiper. Two more team
members pick it up, but the fourth just can't focus on it.
Suddenly, the whole flock gets up and disappears over the

horizon. "Did you see the Curlew Sand?" the other three ask the fourth. He hesitates. "Let me put it another way," offers the leader. "Did you see the flock fly off?" "Of course," answers the fourth. "Then you saw the Curlew Sand." Undeniable. The rules say "all four members must see the bird". Not that they must be aware of seeing it!

All sorts of problems arise from this 'all four team members must see it' rule. Picture this ('cos it's happened). The team arrives on a bridge overlooking a stream: a guaranteed Grey Wagtail site. You're late and behind schedule. You all want a quick tick and away. But one of the team is desperate for a pee…you dive behind a bush. Meanwhile, the wagtail zips under the bridge and off downstream. Three see it, one doesn't. You wait for 10 more minutes. No wagtail. "OK," announces the leader gravely. "We three are going to go for a pee now. You stay on the bridge. I think you know what you have to do." Of course you did. While they were away, you saw the wagtail. Or did you?

And there's the 'I'll take your word for it' situation. In the woods. An unfamiliar snatch of twittering. Just once. No repeats. "What was that then?" three of you ask. "Bullfinch song," comes the answer from number four. "Oh…OK." Well, come on, own up, how many of us know what the dickens a Bullfinch's song sounds like? So what to do? Bow to greater authority, of course. (And banish all thoughts of Great Tits.)

And how about escapes and hybrids? All those dodgy wildfowl? OK, we know you can't count Bahama Pintail, or indeed Barnacle Goose, when they swim over for a nibble at your chocolate Hobnobs. But how about that Pochard x Ferruginous hybrid? You haven't had either species on the day. So what do you do? Count one or the other…or both?!

Or there's even the 'after the event' dilemma. It's getting dark as you scan a flock of about 50 terns over the gravel pit. You suspect Arctics among the Commons, but can't nail any. As you drive away, you call up Birdline South East on the mobile: "Gravel pits: 45 Common Terns and five Arctics"…

Oh dear, I really must stop this. I'm shattering illusions, upsetting people, and bringing back all sorts of memories…oops, what a giveaway! And how about you? Did any of the above bring a little flush of colour to your cheeks, eh? It's OK. Banish that guilt. You've got lots of excuses: you were exhausted, under stress, you wanted to put those smug gits from the bird club in their place, but best of all, just keep telling yourself this — it's all for a very good cause!

Birdwatch bird race
Edited high — and low — lights

The Portland Heights Hotel is a brilliant place for a birdwatcher to stay. At the top of the hill, the gateway to the Bill, the observatory two miles down the road, bushes in the grounds for migrants, and a simply stunning view back over Chesil Beach and The Fleet. It seemed an awful pity to be leaving it at 2.30 am in pitch darkness. But that's what you do on a bird race.

Saturday 16 May. I was part of the *Birdwatch* team committed (or do I mean condemned?) to tazz round Dorset for a day, and a large chunk of the night. Rather like the Portland terrain, bird races are full of ups and downs, pits and troughs, highs and lows. Within the first few minutes we had experienced both extremes. Is there anything nastier than being wrenched from a deep sleep at 2.00 am by a night porter battering on your door loud enough to wake the dead, let alone me and my team mates? But then you have to suffer that in order to experience the undeniable excitement — anticipation, exhilaration almost — of tiptoeing outside and feeling the waft of an east wind on your cheeks, under a star-studded sky and a silvery moon. And that's the last romantic image you'll probably have all day!

The hopes, disappointments and stress start instantly. In the hotel car park, three of us listen for nocturnal bird calls. Maybe a Grey Plover or a Whimbrel flying over. We hear nothing. A pity, but at the same time almost a relief because we couldn't 'count' it since our fourth team member isn't with us yet. So we start worrying. Has he overslept? Worse still, has his car broken down? Two more minutes of tension, then the rumble of an engine, the sweep of headlights, and we are legalised by the arrival of the fourth and undoubtedly most vital member of our team: Dr George Green, local expert and driver.

And we're off. Away from the Bill, which seems instantly ironical, since it is arguably the most famous birding spot in the county, back through Weymouth and Wareham, heading for our dawn chorus site, and praying that a Barn Owl will ghost across the road. It doesn't. Which is probably very wise of it, since George's vehicle seems as accurately programmed as an Exocet missile when it comes to zapping wildlife. Heaven knows he tried to swerve round the rabbit, and clear the hedgehog, but ominous clunks are beginning to conjure up an image of a radiator plastered with road kills. Is he going to mow down the birds like this? We fantasise about ending the day by skidding back to the observatory able to corroborate our final total by pointing at a car festooned with trophies. We don't just count 'em, we squash 'em and mount 'em! Sorry, bad taste I know, but in my experience bird race banter often gets a bit sick. Or is it just me? No. As it happens, indisputably the worst offender in our car was your very own Editor. Shocked? Naw. You've got to have a giggle. It's a bird race tradition.

And talking of sick traditions, an almost obligatory ritual in the early conversation is the swapping of ailments. It is almost as if you are setting up your excuses in case you end up with a lousy score. Thus our team and our handicaps were as follows. Dominic Mitchell: been off work for nearly a week with a mystery virus picked up in Florida. Feeling better now, but desperately short of sleep, and therefore liable to nod off at any moment and become unrousable. Roy Beddard: seemingly fit enough, but has to stop at periodic intervals to take mysterious pills, so he's surely not so well as he looks. Me: suffering from heel blisters acquired on the previous days 'recce', actually twitching a Woodchat Shrike at Portland, and therefore liable to limp and whinge a lot. George Green: certainly a lot livelier than the last time I raced with him a few years ago, when he had been up all night vomiting and dozed all day in the back of the car, but has had even less sleep at his home than we'd had at the hotel. Two hours to be precise! 'Will you really be OK to drive?' we asked nervously. 'Oh, yes,' he assured us. 'It was two hours of really good sleep!' 'Oh, well, that's alright then!'

And indeed it was. George not only drove all day, he also navigated round a couple of dozen sites, never got lost, kept his temper in traffic jams, and kept up an almost continuous running commentary for almost 20 hours, and — in case you were worrying about it — he didn't run over any birds either. What's more, I didn't see him yawn once. I don't know what he's on, but I want some.

But what of the birds? Well, if you've ever done a bird race, you'll probably agree that many of the memories rapidly becomes, understandably, a bit of a blur. But, like I said, you do experience highs and lows. Here's a few of ours.

Best bit of luck: The Little Owl that flew off a roadside fence. Lucky for us and lucky for it, since it flew away from the road and thus avoided George's lethal front bumper. Most efficient tick: Corn Bunting on the Dorchester by-pass at 60 miles an hour (us, not the bird). Biggest act of faith: Crossbill that called once somewhere in Wareham Forest. Dodgiest credentials: a male Goldeneye that has been resident at Radipole for several years now. Most irritating dip out: Lesser Whitethroat. Especially at it was almost the first bird we heard singing outside the Portland Heights next morning. Least impressive rarity: Silhouetted Terek Sandpiper half a mile away at Stanpit Marsh. Best birding moments: pre-dawn on the heath, with Nightjars perched, hawking and churring, and a Woodcock roding straight past us. A pair of Kingfishers posing on an overhanging branch, instead of just hurtling away round a corner like they usually do. Three Yellow Wagtails dropping in at dusk at the Bill, long after we'd given up on them. Grasshopper Warblers reeling at sunset.

Oh, and one more magic moment: 9.50 pm when we finally decided to give up, vowing we'd never do it again. Well, not till next year.

The Hong Kong bird race
April, 1990

This was my second visit to Hong Kong, but the first didn't count. It was a one day stop-over in September, 26 years ago. I don't remember much about it. I recall stepping out of the plane into air that was more fetid than the scent of a birder's socks after a 24-hour bird race; and I have a dim memory of seeing a bunch of Black Kites circling around, hoping I'd pass out and become their next meal...which I duly did. Pass out, that is. Fortunately, there are some things that even Black Kites find inedible and I was one of them. Thus, when I returned to Hong Kong on 4 April 1990 to join the ICI team, I was definitely a new boy.

It was rather intimidating being a 'foreigner' in a bird race team. At best, you are a bit of a handicap. At worst, a total irritant. If you want to avoid being either, it is absolutely essential that you get to know the common local birds as quickly as possible. I had two and a half days to learn that most squeaks and chirrups are made by Bulbuls, most wistful songs are Magpie Robins, most things that perch on wires are Spotted Doves and most calls that sound like Great Tits...are. (Unless they're Tailor Birds.) I'm not sure whether it was a tribute to my speed of learning, my will to live, or the tolerance of my companions that I avoided being strangled on the day.

In fact the 'recces' had many highlights. I managed to pick out my 'own' Asian Dowitchers and Saunders's Gulls; and Malcolm Goude directed me to the much dreamed of Spoonbilled Sandpiper, which looked even sillier than I'd ever imagined. No wonder they are so rare. Too embarrassed to show themselves I should think. The experience of seeing the bird was only surpassed by the joy of watching Martin Williams wading out 50 yards knee-deep in effluent to try and photograph it. We hadn't the heart to tell him it had buried itself into the mud as soon as it caught sight of his camera. Who can blame it? If you had a hooter like that, would you want people pointing telephoto lenses at it?

By the afternoon of the 6th, I felt I knew Hong Kong's birds and its hot spots pretty well. I was even becoming uperty enough to discuss routes and tactics. Two things were becoming clear. One good, one less so. The fact that Hong Kong's best sites are pretty close to one another would mean we'd spend most of our time in the field, rather than crunched up in the back of a car. This I liked. I'd done a bird race in Australia last October which had involved several two- or three-hour drives, interrupted by the occasional half hour's of birding, and that usually out of the car window!. I don't know why they didn't call it a car rally and have done with it. The downside of the Hong Kong situation was that it seemed inevitable that all of the teams would be visiting the same locations, quite possibly at the same time. This wasn't such a jolly prospect. I suffer from ornithological claustrophobia. Back in the UK, I ring 'Birdline' to find out where the big twitches are so I can avoid them. I had visions of 15 teams trying to squeeze into the hide at Mai Po, all talking in code so they wouldn't know what one another were seeing. In fact, this is probably exactly what did happen on Friday evening. We wouldn't know. We decided we couldn't face it. At least, that was one reason we decided to leave Mai Po till Saturday morning. We also figured that it would be impossible to do the area thoroughly in the 40-odd minutes of daylight that would be left on Friday; unless of course you succumbed to the almost irresistible temptation of identifying everything on the scrape before start time and then, on the strike of six o'clock, just assuming it was all still there. 'Well we haven't seen anything fly away, so they must all still be there, so put 'em down!' Any ears burning? I suspect we'd have done the same.

But we didn't. Instead, at half past five, we were slipping silver dollars into a parking meter outside the KGB. Well I think that's what is was called. At this point, I have to admit that my ageing brain had had enough trouble remembering all the new bird names and I never even began to sort out the places. Anyway, I soon discovered that the KGB — or was it the GBH? — was a sort of cross between a zoo and a biological garden (come to think of it, maybe it was the ZBG) and it was the setting for the most bizarre start I've ever had to a bird race. As the clock struck six, we, and at least three other teams, set about the demanding task of deciding whether various equally tame birds were inside or outside cages. A small flock of Laughing Thrushes was undoubtedly inside, but the cage didn't have a roof, so we could count them! By that token, I reckoned we ought to have been able to count the flamingos, who were equally roofless, but we couldn't, because they couldn't fly. The ability to get airborne qualified the cockatoos and parakeets, but not a passing Boeing 747, though I can't think why not, as it seemed to me it was as much a wild bird as they were! Even the Peking Robin was apparently 'probably an escape', but that didn't matter as long as we didn't see it slip back into its cage to roost.

At seven o'clock the park keeper kicked us out, and the first day was more or less over. We enjoyed a pint and fish and chips, wrapped up in newspaper that was no more authentic than the birds we'd just ticked, and went to bed. I could hardly believe it. In nearly 10 years of bird racing in various parts of the world, never have I officially been allowed to sleep for more than about three hours out of the 24. Yet here I was snuggling under the covers in the Mandarin Hotel at 10 o'clock, and not due to be picked up till quarter to five. Nearly seven hours! I felt so guilty, I stayed up writing notes for, oh, at least five minutes.

At 5.15am we were stumbling up the path at Tie Pin Cow (well, that's what it sounded like!), wishing someone had brought a torch. Collared Scops Owls hooted and a Barred Owlet whinnied. We waited for the dawn chorus. There wasn't one. Despite this disappointment, at 6.15, we set off into the forest feeling rather smug that we were in there ahead of the other teams. Meanwhile, no doubt they were feeling smug that they'd avoided standing in the woods for an hour hearing virtually nothing. During the next hour, things changed. We heard plenty. But we saw nothing. Most of what we heard where the tantalising 'tseeps' of invisible thrushes. 'They sound like Eye-browed,' I suggested hopefully. It was about all I could say, as Eye-browed was the only one of the available species I'd heard (in Thailand). Martin put me in my place: 'They all sound like that.' After another half hour failing to get a glimpse, I was getting frustrated enough to moot heresy: 'Well, whatever they are, they are a new species for us...so can't we put something down?' With a withering silence, my companions put me down.

It was perhaps not entirely coincidental that the appearance of several other teams along the forest trails coincided with the appearance of more birds in the treetops. Ironically, one of them was a thrush, which I had to pretend I wasn't getting a great view of. It wasn't an Eye-browed (it was a Pale) but I bet the others were. As we came full circle on our walk, more and more teams appeared. We greeted them with wry smiles and mixed feelings. We'd actually begun to record more and more birds as the day had warmed up, so maybe our rivals had timed it better by arriving now. On the other hand, judging by how many people were in the forest, we were going to have the Mai Po hides virtually to ourselves.

About an hour and a half later we realised why, we were almost too late. In fact we got away with it. The miraculous Mai Po — so wonderful I can even remember its name! — duly offered us nearly all the birds we wanted. Admittedly, the Spoonbilled Sand wasn't willing to be laughed at even for money, but we got nearly everything else. Then suddenly, they were up and away and, before our very eyes, the scrape became totally empty. A quick scamper down the 'boardwalk' confirmed that the tide was way out, and we had been lucky the birds had stayed on their roost so long. If we'd arrived half an hour later, we would have missed the lot!

But we hadn't, and our spirits were high as we cheerily set off on a tour of some of Hong Kong's less salubrious birdy sites. Loch Mac something was less reminiscent of a Scottish beauty spot than of a particularly putrid cesspit, and yet on our recce it had produced half a dozen each of Fantail and Pintail Snipe. We later learnt that, on the race day, several teams had both these species plus Swinhoe's, but by the time we got there no doubt all snipe had got fed up with being disturbed and argued about and, suffering identity crises, had either migrated or committed suicide. We saw a single unambiguous Fantail. Fortunately, other birds were suitably embarrassed by this uncooperative behaviour; a Francolin shouted to us from the hillside, and an exhibitionist Bluethroat posed so long on the telegraph wires that we had to shoo it away in case another team arrived.

We then moved on to what was fast becoming my least favourite 'hot spot'. Lamb Chop valley had been pretty unimpressive on our recce but it was worse on the day. It was hot and noisy, and it seemed drearily appropriate that the only new species it managed to produce was a tatty little Olive-backed Pipit (a Hong Kong 'trash bird' if ever there was one). We were finally seen off by a ravenous black dog who was intent on tasting my newly exposed knees. But if I thought that was unimpressive...

At least I remember the name of the 'Golden Triangle'. It is instantly recognisable as a typical good Hong Kong bird site in that it resembles something between a compost heap and a scrap yard. Apparently, it can be brilliant for warblers and flycatchers. It was the only place we saw nothing new. Nevertheless, we were not down-hearted. There were still a couple of hours to go, our score was 137, and there were several species we were pretty confident we would find at our final destination.

I shall never forget 'The Fence'. For a start, it's a jolly good name: there's a ruddy great fence there. Secondly, this was the scene of my first Spoonbilled Sand. And thirdly, well I'm coming to that. During the next hour, we cleaned up various relatively easy birds: Cormorant, Yellow-breasted Bunting, Pratincole and so on, and one not quite so easy one, Blyth's Pipit. And that was that.

And so we set off back to...Hang on! Wait a minute! Hold it!. Oh, alright. Since our Blyth's Pipit has no doubt probably by now attained the credibility level of Maradonna's 'Hand of God' goal, I suppose I'd better defend it whilst I have the chance. Now, in case you don't know, the identification of Blyth's Pipit is considered something of a challenge; largely because nobody seems to know how to recognise it! There have been various erudite papers on the subject, but basically they boil down to the suggestion that the incredibly rare Blyth's is almost identical to the incredibly common Richard's, except that they are not the same. Exactly how they differ is a matter of some controversy. It seems likely that Blyth's has very slightly different shaped centres to the medium coverts (but only when it's in adult plumage). Now, if you're not sure where the median coverts are (and there's no shame in that), they are a row of little round feathers on the wing, that are at just the right height to be obscured if a pipit-sized bird is creeping around in the grass.

Naturally, 'our' bird was creeping around in the grass. In fact, Martin spotted it first. 'Is that a bunting?' he suggested. I 'scoped it. 'No...its a pipit,' We all 'scoped it. 'It's not a Red-throated'. 'And it's not a Richard's'. Everyone agreed, but why not? 'Well, it's too small...and the bill's too fine...and it's...well, it's just not.'

Then, Martin and I spoke simultaneously. As one, we uttered the blasphemous word: 'Blyth's!' Our confidence was, of course, totally unjustified, since neither of us had ever seen the species! At this moment, the bird flew and landed on the nearby oystershell field. For a tantalising few seconds I had it in the 'scope with it's back to me. I slipped my brain into 'feather check' mode: 'Moderately streaked mantle. Unstreaked rump. Tertials covering wing-tips.' The bird shuffled and showed me it's right wing. 'Median coverts...median coverts...where the *** are its median coverts?' It wasn't that I'd forgotten my feather topography. The fact was...it didn't seem to have any! Were they covered by ruffled feathers or had they moulted, or?

It shifted again and showed me the left side: 'Aha, two big dark centres on the medians. OK, so exactly what shape are they?' Too late. Time's up. Off it went again. This time it landed on a bund about a 100 yards away. Flight: pretty level, no bounding like Richard's. Call? Martin thought he'd heard 'a soft Richard's-like sound.' I'd heard nothing. We broke the cardinal rule of bird racing, and started birdwatching! Down the bund we scampered, nearly blinded by midges, till we realised it was more sensible to scuttle along the lake shore. We crept up onto the bank. There was the Pipit: nipping around catching flies in the grass, which was just long enough to — you've guessed it — cover the medians! Maybe that's why they're called 'coverts?' Another five minutes of mixed excitement and frustration and a glance at the watch. Getting on for five o'clock. We were going to have to leave it. 'Let's at least see if it would call when we flush it.' It didn't. Instead, it disappeared into the haze, leaving us to 'make your mind up time.' By a sort of process of elimination, surely it had to be a Blyth's. But could we prove it? Frankly, no. It was then, and only then — honestly! — that Martin muttered: 'It must be the same bird.' 'The same bird as what?' I asked. 'As the Blyth's that was claimed last week. First for Hong Kong.' 'Where was that seen?' 'Oh, er...on that bund I think.' At least 50 yards away! The coincidence was surely irresistable. So what were we to do? 'Put it on the list and let the adjudicator decide.' Wouldn't you?

Well, the adjudicator accepted it. Whether this was because he'd seen the previous week's Blyth's himself and our description tallied; or whether he felt a slight sense of debt to me 'cos I'd directed him to an Olive-backed Pipit in Yorkshire a couple of years ago (where it's definitely not a trash bird) only he can say (but he probably won't!). Or maybe he gave use the benefit of any doubt in the interests of raising more money. Quite right too. Come to think of it, I wonder if the Blyth's would have been so coy about flashing its medians if it had known there were nearly a 100 dollars riding on it.

So that was that. A couple of Grey Starlings hurtling over in the gathering gloom gave us our 151st species, as we left the Fence and zoomed off in the direction of the nearest traffic jam. And there we stayed for the best part of an hour. As the minutes ticked by, I was full of admiration for Martin's self control, until I realised he was fast asleep. He woke up at one minute to six. His first words were: '*****, *****, and ******'. Which, roughly translated, means: 'Oh dear, we're going to be late, and we lose a species for every five minutes. That's a bit of a shame isn't it. Still, never mind.'

We didn't have to. As we skidded into the Jockey Club gates we narrowly missed clobbering a large dog. We lost valuable seconds considering whether we ought to go back and drape it over the radiator to prove we'd had problems, but we still clocked in just short of five past. We'd lost nothing except our cool. In fact, several other teams were behind us and I trust they weren't penalised either. After all, it's only a game!

The sporting spirit continued at the awards ceremony, as friendly rivals jokingly spat blood and playfully hissed 'string!' at Malcolm's announcement of 'Blyth's Pipit' as our bird of the day. We, for our part, were absolutely thrilled that three or four other teams had beaten our score. Mind you, as we drove back to Central, we couldn't help but think maybe it would have been nice to have won. Especially by one bird.

BITS AND PIECES

And finally...a miscellany of pieces that didn't quite seem to fit under any of the previous headings.

The good hide guide
The pros and cons of birding from a box

I recently opened a suburban nature reserve. The site was an ex-rubbish tip which presumably had been so polluted with leaking batteries, paint stripper and other noxious substances that it had been declared terminally unfit for human habitation. So they'd given it to nature. Years of voluntary toil had cleaned it up and created a perfectly good habitat, complete with wildflower meadow, butterfly bush, a potential wood (at the moment merely a couple of dozen saplings in plastic sheaths) and two small lakes. I was delighted to encourage such effort and enterprise. The opening ceremony was very enjoyable, but then came the words I always dread: the man from the council announced: "Mr Oddie will now lead a bird walk."

My group consisted of a dozen youngsters — who looked and behaved like extras from "Grange Hill" — a harassed teacher, a mayoress in a shiny chain, a 'sponsor' who'd forgotten his wellies, two press photographers and a complete prat from the local radio. The prat led us to the centrepiece of the reserve, a brand new wooden hide. He insisted that we all crammed inside and did a 'live' broadcast'.

"Hi! I'm standing inside a hide," he yelled, in statutory Chris Evans fashion, "along with twitcher Bill Oddie. And a bunch of kids." At which the kids did their impression of the audience on The Word...only twice as loud. "So, Bill, tell us what we can see from in here?" I gave him an honest reply: "Bugger all."

Under the circumstances, it was hardly surprising that there were no birds to be seen that day, but my suspicion is that there never would be. The hide had been built at the end of a promontory with a shale path leading to it, with no screening whatsoever. It was as if the council were so proud of it they wanted it to be as conspicuous as possible. There was no way you could get into it without scaring everything away. And yet they had coughed up a few thousand quid to build it.

The reserve warden agreed that it was a waste of money but he also confirmed that, when applying for funds, a bird hide was one of the likeliest projects to find favour. It's true. Believe me, if you are having trouble getting planning permission to build a new gazebo or garage, tell them it's a bird hide and you'll be OK. They'll probably give you a grant, and get me to come and open it.

Of course, that council hide will have its uses. It will no doubt temporarily reduce vandalism on the nearby estate by distracting the hoodlums who will not rest till they've finally burnt it down. And, until that happens, it will also provide a dark and cosy venue for all kinds of sexual experimentation. Ask any warden of a surburban nature reserve and they'll tell you: all hides might as well be windowless, built of reinforced concrete and fitted with condom machines.

But I mustn't be so cynical (even if it's fun). There are, of course, many bird reserves that have excellent hides. But I still don't much enjoy being inside them. For a kick-off, they are supposed to provide shelter from the elements, but — on a sunny day — it's often as freezing as a fridge in there. And when will they design something you can use a telescope in? A whole industry has evolved to deal with the problem, but it has never been solved. How many of us have a collection of devices, made out of a mysterious metal twice as heavy as lead, and covered in enough screws and clamps to excite the Marquis de Sade? I've seen such things in the London Dungeon recently, but not in a hide. Most people still struggle with a tripod — usually jabbing themselves and others in sensitive areas in the process, or — worse still — blocking one another's view. Just the other week,

I heard of a situation at a twitch where things got so tense and tangled that one birder actually attacking another with a Velbon. The first recorded instance of 'hide rage'? Is that why they've started wrapping up the legs in foam rubber? Before someone gets really hurt.

However, the fact remains that there are some reserves where it is almost impossible to see any birds unless you go in a hide. And it is also true that some people don't seem to notice birds until they are in a hide. I recall being in a hide at Hauxley when a family came in and started scanning through the species on the lagoon. "There's an Oystercatcher...and a Turnstone...and — Oh, look! — a Golden Plover. Wow!"

One Goldie and they were thrilled. Apparently they had been totally oblivious to the 300 they'd flushed from the beach on their way in. Still, I suppose it's a good thing if hides do focus people on to birds. But why do they? Maybe it's because we're conditioned to only believe what we see through a rectangular opening — just like on the telly.

OK then, much as I dislike being inside a wooden box, I have to concede that you do see birds from hides. But you also miss them. One lovely sunny morning this May I was at Minsmere. As I walked across the reserve I spotted five dots approaching from over Sizewell. Bins up. Five Spoonbills. They flew ever closer, circled quite low over the scrape and then carried on past Dunwich and away. A wonderful sight. I truly wished I'd had someone to share it with. At that moment I was passing the West Hide. Two birders emerged. "Weren't those Spoonbills fantastic?" I enthused. They looked positively crestfallen. "Spoonbills? What Spoonbills?" Part of me felt sorry for them. The other part thought: "Serves you right. You really should get out more!"

Love it or hate it

Seawatching

In life there are some things you never forget how to do. Seawatching isn't one of them, as I was reminded this September. I had just finished doing a mini-tour of the West Country, giving slide shows to RSPB groups. On the Sunday came my reward: a day out birding in Cornwall. There was a brisk north-westerly wind blowing. Heaven knows, I would have felt safer combing the woody valleys and seeing nothing — I have lots of practice at that on Hampstead Heath — but, be honest, this was clearly a day to go seawatching.

Thus it was that at about 9am I, and a couple of equally out-of-practice local birders, took our places among the 20 or so seawatchers already cowering under the wall below the Pendeen lighthouse. Being in such a group of course immediately introduces the possibility of embarrassment and the stress of potential public humiliation: you have an audience, and believe me, it's even worse if they know who you are. There was no way I was going to risk speaking until I got my eye in. I soon realised that I was likely to remain silent for quite some time.

As I scanned the sea, my first inward emotion was total panic. Heaven knows, I've done plenty of seawatching in my time, but much of it has been in Ireland where the birds are closer and the crowds are smaller and perhaps therefore less intimidating. For a start, you can hear one another. At Pendeen, whenever anyone 'called' anything it was either carried away on the wind or passed down the line as an increasingly bizarre Chinese whisper. By the time one of them got to me it sounded like: "Dead streetwalker flying left". On reflection, this seemed more likely to have been "Med Shearwater". Never mind that, I had yet to find a plain old Manxie.

Then I realised that my eyes were still in 'Irish distance' mode. The birds weren't passing a few hundred yards offshore, but at least half a mile away. What I'd thought might be a Leach's Storm-petrel was in fact a juvenile Gannet and that possible phalarope bobbing in a trough was a winter plumage Guillemot. Or was it in fact a plastic bag? At that moment a bloke near me announced that that was exactly what it was. Take my advice, anyone bold enough to yell out "Plastic bag!" on a seawatch is worth listening to. So I listened. I followed the calls of the 'expert' and tried to ignore most of the others. After a couple of hours, my eye was back in, at least to the extent that I was fairly sure that several of the calls of "Sooties!" referred in fact to particularly dark 'dead streetwalkers'.

As it turned out, 'Meds' were to feature rather more satisfyingly later in the day. In mid-afternoon the wind was still brisk and the valleys definitely birdless, so we decided to try the sea again. This time at St Ives. This was altogether a more relaxing experience than Pendeen. There was a nice little sheltered balcony where the small gaggle of watchers could actually see and hear one another. The local experts soon identified themselves by displaying a nice line in sarcasm, no doubt born of years of listening to stringers.

I'd learnt the lessons of the morning and remembered that my 20x lens was a lot better suited to seawatching than my zoom. I'd also found my voice. After a few minutes getting the range of the passing birds (closer than Pendeen, as it happened) I was confidently calling Meds and indeed enjoying a brilliant comparison as both Med and Manxie sat on the water side by side. (The same species? Yes, and I'm King Kong.) I even called a juv Pom Skua, which was immediately confirmed by the experts. Surely I was now back in practice, confidence regained. Or was it?

I picked up a distant petrel and called it: "Petrel passing the buoy". I was just about to say: "Er...I'm not sure about it," when the adjacent voice of authority announced: "Another Leach's". Whereupon everyone carried on scanning for something better. Except me. I kept my scope on the petrel. OK, it was battling into a gale, but wasn't it hugging the waves a bit more than the previous Leach's? And didn't it have rather a lot of white on the rump? Then I heard someone ask: "Have you ever seen Madeiran?" "Only the one here," came the reply (referring to a famous past glory of St Ives). "What are they like?" "Well, browner," came the response. No-one was looking at 'my bird'. But I was. "Er, doesn't that look browner?" I thought. And the flight...the rump. But I didn't speak. I watched the bird disappear round the headland and I said nothing.

Which was probably just as well. In fact, I'm absolutely sure that all my thoughts were but the ravings of the incompetent (well, out of practice, at least) mixed with a little auto-suggestion from the voices next door. That bird was a Leach's. But, although I was sort of glad I'd kept quiet, I was also rather ashamed. I'd broken the number one rule of seawatching: never be afraid to speak up.

Anyway, I was punished. Later that day I learnt that — barely half an hour after we'd left St Ives — a Wilson's Storm-petrel was 'claimed'. Ironically, I really could have helped there. I'd seen hundreds of Wilson's off Cape Cod only a week or so earlier. It was the one species I wasn't out of practice with!

Out for the count
Playing the numbers game

So…it's early August. I'm on Tresco, in the Isles of Scilly. Family holiday. It's mid-morning and it's a lovely sunny day, but my wife is still having a lie-in, and the kids are glued to The Big Breakfast. It's certainly been well worth re-mortgaging the house so they can enjoy the healthy benefits of an idyllic outdoor week on one of the world's loveliest islands! Never mind, at least I can enjoy a few hours totally guilt-free birding. So there I am wandering around on the North Downs, counting Wheatears, one…two…er, three…it's quite a challenging task actually. On the face of it, the Downs are fairly flat and easy to scan, but in fact Wheatears are brilliant at hiding behind the tiniest rock, or nipping over the horizon and popping up again 50 yards further on. It'd be all too easy to count the same one twice. Three…four…five…or is it just two flitting ahead of me? Start again. One…two…oh, it is five. All in the same binful. Damn, a dog's put the lot up. Now there's none. OK. Wait. Scan across. Aha, there they are. Way to the left. One…two…wait a minute, that's a moulting male (as opposed to the Isabelline look-alikes I've been seeing before.) So that must be six…or is this a completely new group? Scan back right. There's one still there…and another…. Wow, this is really tricky. OK, deep breath, start again. One…two…three…. As I count, slowly and insidiously a question weedles its way in between the numbers: why am I doing this? Counting them, I mean? In fact, why do I count every single bird I ever see? And why can I never rest easy until I have an accurate double-checked total for each species in my note book? 'Get a life,' a voice taunts me. Oh come on, be fair, I protest, every birder counts everything. Don't they? Well, do you? I do. Always have.

Maybe its a generational thing. I suspect it started for me way back in the 1950s, when I took on the taxing responsibility of doing the official wildfowl counts at my local reservoir near Birmingham. I don't even know if wildfowl counts still exist these days, but back then, on a certain day every month, birdwatchers all over the land went out to their particular patch of water and counted whatever ducks, grebes, geese etc were swimming thereon. The whole thing was organised and co-ordinated by the then easily sayable Wildfowl Trust (now become that well known tongue-twister that WWT stands for). Being an authorised wildfowl counter made you feel terribly important. Especially if you were an early-teen schoolboy like me. It was indeed a great responsibility to entrust to one so young, but, looking back, maybe it was no accident that I was granted the counting franchise to Bartley Reservoir, a bleak concrete-sided expanse of grey and lifeless water that very rarely attracted any wildfowl at all. Maybe they figured I was unlikely to make many mathematical or identification errors, because there was usually bugger all there. Not that that stopped me counting it. Many was the form I returned duly and neatly filled in with a 'nil' besides every species. What's more, I convinced myself that my data was every bit as valuable as the counts of massive flocks returned from other Midlands reservoirs.

Alas, it was these flocks that tempted me into a disgraceful piece of behaviour. One weekend I made the mistake of cycling over to Bittell Reservoir, where I experienced what it was like to count more than 'nil' wildfowl. My whoops of euphoria scared most of the birds off the water. They circled round me, then — instead of landing again — they flew away to the east. I cycled back to Bartley, only to find all the Bittell ducks landing on 'my' reservoir, whereupon I naturally couldn't resist counting them. Neither, from that day on, could I resist biking over to Bittell on every wildfowl count weekend and scaring the birds into making the crossing to Bartley. It is possible that the nice man at the Wildfowl Trust noticed the suspicious increase in Bartley's counts and the oddly identical figures from Bittell, but had sympathy and chose not to have me struck off the counters' registry. However, if he didn't suss me, I hearby confess, and apologise if this means that the entire Wildfowl Count database from the 1950s has now to be revised because of my delinquency.

Anyway, if wildfowl counting started my obsession with numbers, the West Midland Bird Club turned it into a serious addition. I discovered that I could rack up a vast total of initials in the prestigious Annual Report by counting all sorts of birds no one else could be bothered with. Maximum gull roosts, numbers of Skylarks flying south-west in two hours, breeding populations of House Sparrows, and even complete absences: 'Turtle Dove — none reported at Bartley this year'. Every credit to W.E.O. (That's me, that is) fuelled my egotism.

Bird observatories made it worse, especially Fair Isle. I first went there when I was about 16. The temptation to show off at the evening log-call was utterly irresistible. (No doubt it still is.) This is how you do it. The warden calls out for a species 'Spotted Flycatcher'…and everyone chips in with their numbers. 'Four…six…any advance on six? 'Seven' 'No more?' OK. Seven it is.' Then, just as the warden is writing 'seven' in the Spotted Fly column, you finish adding up your total and announce: 'Nine!' Since he doesn't want to have to alter the number, the warden is likely to challenge you. 'Are you sure there weren't any overlaps?' (That is, maybe you counted one bird nine times.) Naturally you are so affronted by this accusation that you begin to reel off exact times and places of each sighting, and indeed claim that you could recognise individual bird's distinguishing marks. This is so tedious that he'll soon go 'Yes, yes OK', and will accept your 'nine' — and probably all your other counts too. Well, so you think until you read the log the next morning. Why do you suppose wardens always write in pencil? (Figures easily rubbed out later…OK?)

But still the counting has continued. Even abroad, in faraway places unlikely ever to be revisited either by me or any other birder; which surely makes the exercise fairly pointless. From my 1981 Thailand notes: 'In a paddyfield, somewhere between Ping Pong and Tie Pin — 137 Open-billed Storks,' So?

Of course there are valid reasons for the counting habit. It adds shape and focus to your birding. It makes you sort through flocks, and gives you a better chance of coming across a rarity. It sort of dignifies your note books, and indeed your statistics might even be useful, to the BTO or somebody. But of course, above all else, the real reason I count everything is to try and distract from the fact that, at least part of the time, birding can be incredibly boring! Which take me back to Tresco in early August. Now where was I? Five…seven…eight…twelve Wheatears! Wow — almost an invasion. Ring Birdline immediately.

Life on Mars

Diets for birders

Man — or woman — cannot live on birds alone. Even birders have to eat and drink now and again. However, a birdwatcher's sustenance regime is surely rather different from that of normal people. Or is it just me?

For example, take breakfast. It's true I do know some birders whose eyes simply do not focus until they have been fuelled by an intake of half a gallon of steaming brown brew and a couple of kilos of cornflakes, but, personally, I am the exact opposite. If I'm on a birding trip, I rise at dawn and nothing passes my lips except a tooth brush until I have flogged every bush and burnt up every migrant within a half-mile radius. It's not that I need to work up an appetite. My motives for depriving myself involve elements of fear, competitiveness and masochism. The fear comes from my conviction that not only does the earliest bird catch the juiciest worm, the earliest birdwatcher will find the rarest bird; and if I am not the first out, it won't be me. Which brings me to the competitiveness. There's nothing I relish more than being able to return to the observatory, B & B, tent or wherever, and grip off the late risers. I stand back to avoid the shower of spluttered rice crispies and make sure I mutter 'Its probably gone by now', thus ensuring maximum guilt, panic and indigestion. Very satisfying. What's more, there'll probably be lots of abandoned hot tea and toast, so I don't have to bother to make my own, and it will, of course, taste all the better because it will be a sort of reward for finding the mega-gripper on an empty stomach. You see, in my opinion, you have to earn breakfast.

Which is where the masochism comes in. This certainly derives from my early local-patch days near Birmingham, when I used to trudge round a bleak and birdless reservoir every weekend. Masochism incarnate in itself, and made all the more agonising by hunger pains. But therein also lay the motivation. My incentive for completing the three-mile circuit was that it finally led to a truly wondrous 'greasy spoon' transport café in the nearby village, where — in an atmosphere as fetid as a trucker's armpit — I would consume a bacon butty with bread as thick as house bricks, lubricated by half a pound of real butter and lashings of Daddy's sauce, and washed down with a giant mug of frothy coffee resembling something you'd see bubbling in the crater of a volcano. (Oh, they don't make 'em like that any more.)

My present day local patch is Hampstead Heath, where I still practise the breakfast/reward principle. My intake must not exceed three extra-strong mints, a banana and a couple of Rennies until I have done the circuit and thus qualified for a bagel and smoked salmon at the local tea rooms (they don't do bacon butties in Hampstead).

Further elements of my birding/eating habits were formulated during various visits to Shetland back in the Sixties and Seventies. It was on Fair Isle that I learnt never to eat a proper lunch. Anyone who has visited the magic isle will confirm that it is bigger than it looks. The observatory building is in the north, but the best bird area — the crops — are in the south. At an average birding amble you just about get to the beginning of the good part by late morning. At which point, if you want a proper lunch, you have to walk all the way back to the obs. Then, after the midday meal, you set off south again, until you have to turn round and come back for tea. It is perfectly possible to spend a week on Fair Isle and never get to the south! You'll have eaten some great food, but you may well also have missed a lot of good birds. Thus I discovered the Fair Isle shop, which just happens to be conveniently halfway down the island, and there I purchased my first birding picnic lunch: a block of cheddar cheese, a bag full of tomatoes, and a Mars bar. It is a menu I have never seen reason to vary over the course of 30 odd years of outdoor dining.

The Mars bar of course deserves a commemorative monument of its own. The nearest I've come to awarding it one was on the island of Out Skerries, where my birding companion of those days and I named a geographical feature in its honour. We discovered that — rather the reverse of Fair Isle — Skerries had a good area of crops situated just in front of our chalet, but a few birds were inconsiderate enough to favour a sheltered geo (Shetland for 'cleft in the cliff') right down the south of the island. It was a long walk, and all too often it produced nothing worth the trek. However, we were too knackered to just turn round and plod straight back, and thus we needed an excuse for a rest. It was there that I developed my undeniably disgusting but very protracted technique for consuming a Mars bar. This consists not of biting and munching, but of licking it very very slowly, teasing the layers into different shapes with my tongue, and finally wiping my lips with the wrapping paper and slurping in the final vestige of melted chocolate. Believe me, this way a Mars bar can be made to last up to half an hour, during which time you will either have spotted a rare migrant sheltering on the cliffs, or have been transported into an erotic day dream (featuring the unlikely casting of you as a Mars bar and Michelle Pfeiffer as a birder. Lady birders recast to suite your taste, as it were). But I digress. Anyway, that spot on Skerries will forever be known in our books not as Trolli Geo but Mars Bar Gulley. And indeed, many's the good bird we saw by lingering there. Which brings me to the nub of my thesis.

The fact is, I reckon that food and drink are as integral to the overall birding experience as the birds themselves. Certainly, I can re-read my note books and not only mentally visualise what I saw, but also taste what I was consuming at the time! I confirmed this when I was browsing through the fascinating Biographies For Birdwaters the other day and found myself playing a word association game, as each venerable name conjured up not only a species, but also a culinary experience. For example: Charles Bonaparte...Gull...Cheese and tomatoes. 'Cos I found a Bonaparte's Gull whilst lunching at Farlington Marshes years ago. Peter Pallas...Warbler and Fish Eagle...chicken curry. 'Cos I saw both species one evening at Bharatpur in India. Edward Sabine...Gull...diced carrots. 'Cos I'll never forget that pelagic. And, best of all, Alexander Wilson...Storm-petrel and Phalarope...fish and chips and Irish whiskey, with which we celebrated a fantastic September day in County Wexford when we had a Wilson's petrel from the ferry, and two Wilson's Phalaropes at Tacumshin Lake, along with three Baird's, two Buff-breasts, four Pecs, and a Semi P! And I'll drink to that.

Owed to a nightingale

Syllables for songwriters

I recently attended the launch of the British Trust for Ornithology's Nightingale appeal. The event was sumptuous almost to the point of surrealism, and, if I may say so without meaning any offence, hardly typical BTO, since it had all the ingredients of a full-blown, gimmicky, media-grabbing publicity stunt. Excellent stuff, in fact. It was held in the middle of London in Berkeley Square, pause whilst you work that one out. Oh, come on. Remember the old song A Nightingale Sang in Berkeley Square? Nice one. Even nicer was the fact that we were all sheltered from the unseasonal hailstorm outside by being under the roof of the West End showroom of Britain's leading Rolls Royce dealer. The manager must have been having kittens at the sight of scruffy, anoraked birdwatchers leaning on his Silver Clouds. But I couldn't resist it. I thought if I could get myself pictured on the evening news that way they might give me a free one. (Rolls that is, not anorak.) In this shrine to opulence, the BTO distributed copies of the bird in questions latest release: a CD of The Nightingale's Greatest Hits, a splendid compilation of poems, cello pieces, World War Two bombers, and over a dozen different Nightingales, singing in a variety of regional accents. All this plus 'celebrity guests', live and in person, and top of the bill — no, not an egocentric pun — was none other than the truly legendary and eternally charming Dame Vera Lynn. Strictly speaking, I suppose they may have got the wrong person, since, as I recall, Vera Lynn's unforgettable wartime, troop-rousing hit was in fact There'll be Bluebirds over the White Cliffs of Dover, not A Nightingale Sang in Berkeley Square. But it doesn't really matter. They are both great songs, and both are equally ornithologically unlikely. An

Eastern Bluebird at St Margaret's? I don't think even Lee Evans would go for that. And a Nightingale singing in a 20-metre rectangle of manicured grass, with a dozen 40-foot plane trees? Fat chance. On the day of the launch there were two Mallards, 30 pigeons, and a pair of Herring Gulls on a nearby roof; and I don't suppose it ever gets much better than that.

Of course, the famous Berkeley Square Nightingale record was almost certainly a stringy report of some typical urban night singer, such as a Robin, Blackbird or Song Thrush. I pedantically pointed this out at the launch by recounting a tale that has happened to me more than once. Most recently this February, when a very sweet elderly lady neighbour of mine stopped me in the street to tell me that she had had a Nightingale singing in her back garden for several evenings, and would I like to come and see it? I hadn't the heart to disillusion her, so I said" 'Oh, that's lovely. I'll try to pop round when I'm free'. As it happens, I was taken suddenly busy for the next three weeks. Early March, she dropped a note through my front door: 'Dear Mr Oddie, the Nightingale is still singing. It has moved to the front garden. So if you happen to be passing...' I still didn't reply. But she wouldn't give up. One evening at about 10 o'clock she phone me: 'Mr Oddie, the Nightingale is singing right now! It's sitting on my garage roof. If you come to your window you can see it.' And thus the scene: I was at my window with phone in one hand, looking out across the street to where my neighbour was peering out of her window, also phone in hand, beaming blissfully and pointing up at a silhouetted bird singing away on her roof. She grinned at me and gestured: 'Isn't it lovely?'. I smiled back and thought: 'Yes it is. It's also a Blackbird, you silly old bat!' But of course I didn't say it. In fact, I still haven't disillusioned her. Which may be kindly, or cowardly, or possibly unwise, since no doubt she — and thousands like her — continue to dwell in total ignorance of the migration schedules, habitat requirements, limited distribution and rarity of real Nightingales, and blissfully believe that they are in fact incredibly numerous all over the country at all times of the year. So why does the BTO need an appeal?

Oh well, that's their problem. I'm not here to write about serious ornithological issues. I'm here to be silly. Which takes me back to that song. Remember the full lyric: 'I may be right, I may be wrong, but I'm perfectly willing to swear, that when you turned and smiled at me...a Nightingale sang in Berkeley Square.' Well of course they were wrong. 'Cos it was obviously a Robin or a Blackbird. And they probably knew it. But that wouldn't have scanned would it? You try singing it: 'A Ro o obin sang in Berkeley Square', 'a Bla a ackbird sang in...' Doesn't work, does it? You see, the lyricist needed a three-syllable bird. Not so easy, as it happens. Think about it. Go down the checklist. Try singing them. 'A Ptarmigan sang in Berkeley Square.' Not really. 'A Guillemot sang." No, even dafter. Neither of them even sing. 'A Cormorant croaked.' Hardly a romantic image. Are there any three-syllable songbirds at all? Surprisingly few. 'A Treecreeper sang.' It'd never be heard above the sound of the smile, let alone the traffic. 'A Flycatcher...in Berkeley Square.' Call me weird, but that conjures up images of George Michael lurking around. No thanks. 'Willow Tit?' Not exactly famous for its mellifluous melodies. In fact Mistle Thrush is the best I can come up with that has sufficient syllables and also produces a half-decent sound; and, as it happens, is a damn sight more likely in a plane tree in the middle of London. But I fear it doesn't really cut it on the romantic image front, even if Dame Vera sang it as sweetly as she could: 'When you turned and smiled at me...a Mistle Thrush sang in Berkeley Square.' No, might as well be accurate and go for the Herring Gull. Be honest, whoever wrote that song was bloody lucky that the one and only three-syllable songbird on the British list that really could have a number one hit record is indeed Nightingale.

So good luck to the BTO, and their appeal, and their CD. I can't wait to see Chris Mead miming it on Top of the Pops!

Now you see it...

The case of the disappearing lough

I was browsing through some recent reports recently when I noticed: 'Leach's Petrel, Akeragh Lough, County Kerry, Ireland.' 'Oh,' I muttered 'it's still there then it is?' I meant the lough, not the petrel. You see, Akeragh has a habit of coming and going, as I first realised way back in 1970.

How did I come to be there anyway? Well, back in those days, Akeragh was truly famous. It was the local patch of legendary Irish birder Frank King, who turned up a list of American rarities that nowadays you'd only expect from maybe the Azores (or indeed America). Small flocks of American Wigeon, Pecs, dowitchers, yellowlegs etc. etc. They'd all popped in to Akeragh, and therefore, on 9 September 1970, so did I. It was one of Akeragh's off days. Was this really the place, I wondered? I'd followed the instructions: 'Park at Ballyheigue beach, walk past the caravan site, climb up on the sand dunes, and look down back towards the coast road.' All I could see was what looked like a glorified puddle: full of water, but almost devoid of birds. Even so, the dangerous combination of information and optimism cast its spell, as my notes relate. 'Two probable female American Wigeon, seen early and identified by greyish head, salmon flanks etc.' At least by the end of the day they'd acquired square brackets and an additional caution: 'I lose confidence a little from the fact that by evening only one wigeon was visible, and that a female European'.

Always knew I was a stringer eh? Ah, but you have to appreciate two things. Firstly, back in them days, we really didn't know how to identify a lot of these rarities (we'd only just discovered that birds had axillaries, let alone noted what colour they were). And, secondly, Akeragh and its birds really did change from day to day as I discovered to my delight when I returned at the weekend. The area had been transformed. By what circumstances I really don't know, but it now looked custom-built for rarities. The water level had dropped dramatically, there was a muddy shore round the main lagoon, plus several smaller pools, all fringed with reeds. And as was, and to a point still is, typical of the west of Ireland, there were no people, but lots of birds. Dunlin, Ringed Plover, both godwits, Greenshank, Ruff, Curlew Sands. OK, nothing 'special' yet, but, believe me, it sure was exciting tiptoeing through the shallows and anticipating what might be round the next corner. And then there it was: my first Buff-breasted Sandpiper. I confess I'd been scrutinising every Reeve, assuming they would be the 'confusion species', but no problem. As they say: 'when you see it, you know.'

What I wasn't quite so sure about was the 'peep' that scuttled out from behind the Buff-breast. I did remember that Frank King had claimed several Semi-palmated Sandpipers from Akeragh, and I even had a copy of an Irish Bird Report in the car, with a rather distant black and white photo of one of them. Or was it? The fact is that the criteria for sorting out 'Semi-P's really were a bit dodgy back in 1970. This was before Grant and Mullarney's 'New approach'. DIM Wallace had, as ever, produced some natty little paintings and plenty of pioneering theory but, at the end of the day, I suspected that jizz and instinct entered into it a little too much to satisfy a rarities committee. So what did I do with my bird? Well, re-reading my notes, there are two pages of feather woffle (signifying nothing!); a rather desperate comment that the bird 'looked very like the one in the photo,' and an even more pathetic attempt to back it up with circumstantial evidence from the next day, when I saw four Little Stints that 'never struck me as anything other than Little Stints.' But I still didn't tick Semi-P!

Neither did I tick the mystery bird that left me with a very weird feeling on my final evening at Akeragh. I was sploshing through the reeds when I flushed 'something'. What it was, to this day, I do not know. It was a real corner of the eye (or sole of the welly boot) job. Up, down, gone. I saw it — sensed it, more like — for barely a second. My notes and memory record: 'probably a crake.' The only feature I noted was 'pale patches on the wings'. Subsequent research led me to Yellow Rail, a bird which is not only very rare but reputedly invisible even in America! Akeragh was good, but surely not that good. I flushed a more feasible solution in Spain quite recently, as I experienced a definite feeling of déjà vu when I trod on a female Little Bittern. Mmm, maybe. The fact is, I'll never know. The bird is gone for ever. What worries me more is possibly so is Akeragh Lough.

I was in Kerry about five years ago. I went to Akeragh. No water. No birds. Only, with supreme irony, in the middle of a bone-dry field, a sign saying 'Bird reserve.' Is it? If so, does it still exist? I have asked Irish birders and had no satisfactory answer. Then I saw the recent report, and took a little heart. But then I've seen Leach's Petrel flying over a building site. Tell me it isn't so.

In your dreams...
Nocturnal fantasies

Do you ever dream of birds? (Anyone who makes a silly sexist remark will have their bins confiscated. And you're giving your age away, 'cos they're called 'babes' these days. Or so I'm told.) I'll ask again. Do you dream of birds? Yes, I'm sure most birders do. I used to really envy them. You see, until relatively late in life, I swore that I didn't dream at all. About anything. Then — around 15 years ago — I was the 'guinea pig' on a BBC QED science programme about dreams. My dreams. I told them I wasn't the ideal subject, since I didn't dream. However, the director assured me that wasn't possible. "Everyone has dreams," she said. I thought she was about to burst into song like in a Hollywood musical, but unfortunately not. She merely explained that I did have dreams, but I didn't remember them. On the programme, they would make me remember. However, they then came up against a bit of a problem. I couldn't go to sleep. Which wasn't all that surprising, since I had a fistful of electrodes elastoplasted to my head, a microphone strapped to my throat, and a camera up my nostrils. So, for the purpose of the programme, they made up a dream for me, so that a Freudian 'shrink' could analyse it. Since my counterfeit dream was based on my hobbies and my history, it involved my mad mother chasing a Pechora Pipit, while I stood by playing a saxophone. Obviously the dream was a fake. I would have been the one chasing the pipit.

Nevertheless, the QED programme did make my subsequent nights more entertaining. I believed what the telly told me. Everyone does have dreams — even me — and, ever since that day, I have remembered bits and pieces of them. And I do, now and again, dream about birds. I suspect the dreams fall into pretty classic categories. There's the 'frustration' dream, where I'm in pursuit of some little brown job which never gives me a decent view. Or I hear of a rarity just up the road, and I can't find my car keys (or did that really happen?). Then there's the 'wish fulfilment' one. Usually, this involves an impossible mix of familiar locations. So I'm wandering round in the concrete bleakness of Bartley Reservoir, enjoying a fall of Bluethroats and Lanceolated Warblers that my subconscious has transplanted from Fair Isle. Then there are the totally puzzling dreams. Like the one I had last night. (Well, the night before I wrote this.)

It just happened that my wife, Laura, had to be up at six o'clock to go off and do a radio show. Consequently, her alarm clock woke me abruptly in mid-dream. (Pity it didn't wake her.) Maybe that's why the images are so vivid. I was in the kitchen of some kind of bird observatory. I have no idea where, except that it was somewhere in Britain. There were a few birders there, but only one bloke I knew. He wandered outside, leaving me to chat to the others about how few birds there were around that day. Then suddenly a bird swooped past the window. As it veered away, I noted it had a white tail tip. I shouted out: "Wood Swallow." No-one else seemed to appreciate that this would have been a major crippler — a Palearctic first — until the familiar bloke raced back in yelling: "Did you see that?" We both repeated: "Wood Swallow!"

The next thing I knew — that's what people always say when they're recounting a dream, isn't it? — we were outside in a sort of garden. There were birds everywhere. Clearly some kind of massive fall was taking place. The puzzling thing was, though, I didn't recognise any of the birds. Worse still, no-one else did. I kept asking: "Anyone any ideas?" No-one replied. To make the panic complete, I had forgotten to pick up my binoculars, and when I rushed back into the kitchen they were nowhere to be seen. At this point, a woman lent me a small pair of opera glasses. I raced outside again and began trying to get decent views of some of the birds. Amazingly, I did. I even began to make mental descriptions of the plumages (I'd lost my notebook, too). But then I felt myself sinking, under dual tides of ignorance and humanity. More and more birders began arriving. The twitch became frenzied, and I was being crushed against a fence, in a horrendous ornithological equivalent to a football crowd disaster. Then Laura's alarm went and I woke up.

I jumped out of bed and wrote up descriptions of the birds I could remember. There was the Wood Swallow. Then a beige warbler-type thing with orange spots and streaks (a bit like a Yellow Warbler, but it wasn't) and — utterly vivid this one — a large thrush, which was jet black except for a scarlet undertail and throat. Then I repaired to my library, and scoured my world-wide field guides. I can't find any of them anywhere. Except the Wood Swallow. And yes, they do have white tail tips, so I'm counting it.

What's more, I'm starting a 'dream list'. So if anyone recognises the other descriptions, do let me know.